The Literature of Chemistry: Recommended Titles for Undergraduate Chemistry Library Collections

The Literature of Chemistry: Recommended Titles for Undergraduate Chemistry Library Collections

Judith A. Douville

Science & Technology Editor
Choice Magazine

Association of College & Research Libraries
A division of the American Library Association
Chicago 2004

Permission to reprint the "Journal List for Undergraduate Programs" of the American Chemical Society, Committee on Professional Training, which originally appeared on the Society's Web site, is gratefully acknowledged. Reprinted with permission of the American Chemical Society.

The paper used in this publication meets the minimum requirements of American National Standard for Information Sciences-Permanence of Paper for Printed Library Materials, ANSI Z39.48-1992. ∞

Library of Congress Cataloging-in-Publication Data

Douville, Judith A.
The literature of chemistry : recommended titles for undergraduate chemistry library collections / Judith A. Douville ; science & technology editor, Choice magazine.
    p.   cm.
 Includes bibliographical references and indexes.
 ISBN 0-8389-8308-1 (alk. paper)
 1. Chemical libraries—Book lists. I. Choice (Chicago, Ill.) II.
Title.

 Z675.C47D68 2004
 016.54—dc22

                   2004018053

Printed on recycled paper.

Printed in the United States of America.

08 07 06 05 04   5 4 3 2 1

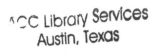

Table of Contents

Chapter 5
Organic Chemistry
75

Chapter 6
Inorganic Chemistry
105

Chapter 7
Environmental Chemistry
120

Chapter 8
Industrial Chemistry
133

Chapter 9
Polymer Chemistry
143

Chapter 10
Biological Chemistry
151

Chapter 11
Internet Resources in Chemistry
164

Journals in Chemistry
169

Index
171

Author's Preface

The *Literature of Chemistry: Recommended Titles for Undergraduate Chemistry Library Collections* is a guide to the most important chemical literature information resources for undergraduate academic collections. It is intended to serve as a collection development tool and a bibliographic instruction resource for librarians, faculty, and the students they serve. It includes annotations for most of the titles listed and useful lists of Internet sites and key journals in the field.

History

The author first became acquainted with the predecessor for this publication in 1965 when teaching at Annhurst College in Woodstock, CT (now closed). A fellow chemistry instructor handed her a thin green pamphlet titled *Guidelines and Recommended Title List for Undergraduate Chemistry Libraries*[1], a report of the Advisory Council on College Chemistry (A.C.C.C.). Its subtitle further described it as "A Report of A.C.C.C. Panel Meetings on Guidelines for Booklists and Library Development...." This 44-page publication, distributed to chemistry departments nationwide, offered a rationale for its creation and some ideas about selection of library materials for chemistry libraries that catered specifically to undergraduate chemistry students.

In 1969 a revision with the same title was issued, but it would be more than a decade before the next revision appeared. A.C.C.C. apparently had disbanded; many of the original chemistry faculty who made up the advisory panel had either retired or died, and as departmental libraries lost favor, less emphasis was placed on creating specific library collections outside the main university or college library. However, the idea persisted that such collections might be useful, especially on larger college and university campuses, and indeed many institutions had and continued to maintain these "special" libraries, often funded by departmental monies rather than by the main campus library.

In 1982 a new revision titled *Guidelines and Suggested Title List for Undergraduate Chemistry Libraries*[2] was published by the American Chemical Society's Division of Chemical Education jointly with ACS's Department of Educational Activities. Student helpers were paid to literally cut and paste citations, chosen by the compiler, onto file cards from which a master list of titles was created. The ACS recruited editors to distribute the list to about 70 college and university reviewers who agreed to rate each title; the ratings were used to choose the "best" items for the list. From this large list a small publication resulted, some 76-pages long, with unannotated, brief author/title/publisher citations only. Journals, abstracting and indexing services, and major treatises were listed. This 1982 revision was the last in this A.C.C.C. family of *Guidelines*, sponsored by the American Chemical Society.

Since the 1982 revision, there has been a void in the literature. The lack of an up-to-date resource to the chemical literature specifically focused on undergraduate collections was the impetus for this current compilation.

Selection of Titles for This Guide

Many of the titles included in this guide had their origin in columns written by the author for the *New England Association of Chemistry Teachers Journal* (*NEACT Journal*) over a period of about 14 years. The columns covered various areas of the chemical literature and were intended to help chemistry teachers and professors cope with the everincreasing bulk of chemical literature, and to help students at all levels find the correct works to answer their chemical questions. Titles discussed in these columns were entered into a database, which was updated as newer editions or more appropriate titles became available. Some of the works in the database were admittedly very old, but among these were irreplaceable resources that should never be purged from library collections. In addition, as other relevant titles became available or new areas of study needed to be represented, the author expanded the database, which over time grew to include more than 2,000 titles. This database formed the basis from which titles were evaluated and selected for inclusion in this guide.

---

[1] **Advisory Council on College Chemistry.** *Guidelines and Recommended Title List for Undergraduate Chemistry Libraries*; a report of A.C.C.C. panel meetings on guidelines for booklists and library development, Washington, DC, February 1965 [and] Crawfordsville, Indiana, October 1965. Stanford, CA: 1966. 44 p. (Its serial publication, no. 12)

[2] **Guidelines and suggested title list for undergraduate chemistry libraries**; edited by Robert C. Brasted and Leallyn B. Clapp, from a list compiled by Judith A. Douville; sponsored by the Division of Chemical Education, ACS, and the Department of Educational Activities, ACS. Rev. American Chemical Society, 1982. 76 p.

To ensure authoritativeness, two levels of outside review were commissioned. Initially, three chemistry or science librarians evaluated the manuscript, and supplied comments about the work, suggestions for new titles and replacements, and other helpful insights. The manuscript was revised to reflect these suggestions and sent for evaluation to five new reviewers, who were selected for their chemistry backgrounds and membership in the Division of Chemical Information of the American Chemical Society. Reviewers rated each title for its overall value for undergraduate students, using the following ratings developed especially for this task: "essential," "highly recommended," "recommended," "optional," "not recommended," and "unfamiliar with this title." Reviewers also rated each title with which they were familiar by user level, ranging from two-year college through graduate level. This guide includes all titles that received a "recommended" or higher rating by at least three evaluators and that were appropriate for any level of undergraduate collection. Titles rated exclusively at the graduate level were omitted. The bulk of the nearly 1,110 titles included in this guide date to 2002; a few 2003 titles are included.

Scope and Organization

This work is organized into ten areas of the chemical literature, beginning with general materials as a standalone brief introduction to the most important basic resources, then treating some general science resources that have a bearing on chemistry. The next chapters treat, in order, analytical, physical, organic, inorganic, environmental, industrial, polymer, and biological chemistry. Each chapter is arranged in subsections reflecting particular types of resources, e.g., dictionaries, encyclopedias. Every effort was made to appropriately place materials based on their use and major content. The final chapter presents a selected list of some fairly important and stable Internet sites. A list of core chemistry journals follows. A combined index lists authors, subjects, and titles.

Most titles on this list are annotated; however, some lack annotations because efforts to examine certain titles or find suitable descriptions in the literature were unsuccessful. Prices are not included because they change readily, and many items are no longer in print. Users of this guide may find newer in-print items for sale from numerous vendors; out of print items may be available electronically on the Internet, or from used booksellers.

Readers will find this new publication much more complete than any predecessors. Internet sites of special interest to chemistry educators and students are believed to be a first for this type of chemistry resource. The major departure from preceding guideline publications, however, is the division of information into the various areas of chemistry. It is hoped that readers will find this arrangement especially useful for evaluating and establishing chemical library collections in colleges and universities worldwide.

Entries are assigned a record number and include the following bibliographic data: authorship, title, edition, place and date of publisher, pagination, ISBN, Library of Congress classification number, Dewey number, and Library of Congress book number.

A typical entry follows:

> **1.026**
> **Hampel, Clifford A. and Gessner G. Hawley.** *Glossary of chemical terms.* 2nd ed. NY: Van Nostrand Reinhold, 1982.
> 306 p. 0442238711 QD5 540 81-11482
> Short paragraphs to explain terms. Some drawings. About 2,000 terms used in chemistry and the process industries. Brief biographical sketches. Cross-references.

The record number signifies that this entry appears in the first chapter. There are two authors, and the work is in the second edition. It was published by Van Nostrand Reinhold in New York City in 1982. The ISBN is 0442238711; the Library of Congress classification is QD5, the Dewey classification is 540 (Chemistry), and the Library of Congress number is 81-11482. There is a brief annotation.

For edited works, the title is the first bibliographic element in the record (the "main entry"); the editor statement follows the title. All other elements of the citation are as noted above.

Acknowledgments

This compilation could not have been prepared without the expertise and the cooperation of the five final reviewers, who evaluated all titles, supplied comments and suggestions, and offered encouragement that served this project admirably:

Jeremy R. Garritano
Chemical Information Specialist and Assistant Professor of Library Science
M. G. Mellon Library of Chemistry
Purdue University

Patricia Kirkwood
Science Librarian and Assistant Professor
Mortvedt Library
Pacific Lutheran University

Alison S. Ricker
Science Librarian
Science Library
Oberlin College

Bruce Slutsky
Senior Science/Engineering Reference Librarian
Robert Van Houten Library
New Jersey Institute of Technology

Leah Solla
Chemistry Librarian
Physical Sciences Library
Cornell University

Active in the initial reviewing phase and an individual who supplied much valuable advice was:

Gary Wiggins
Director, Program in Chemical Informatics
Interim Director, Program in Bioinformatics
Adjunct Professor of Informatics
School of Informatics
Indiana University

About The Author

Judith A. Douville is a graduate of the University of Connecticut (BA, chemistry, 1959; MS, organic chemistry, 1965) and the University of Rhode Island (MLS, 1971). She is the Science & Technology/Art & Architecture Editor, *Choice* Magazine (published by ALA). Before joining *Choice*, she taught chemistry at Annhurst College and chemistry and physical science at Central Connecticut State University and Eastern Connecticut State University; she returned to Central Connecticut State University to teach chemistry part-time after nearly 20 years' absence (spent in the industrial sector). She also served that same university as Serials Librarian until, in 1972, she joined the staff of *Information Science Abstracts* as managing editor; from there she managed the Metals Information Center at Olin Corporation in New Haven, CT; headed the technical library at TRC Environmental Consultants, in East Hartford, CT; served as an industrial consultant at the Chemists' Club Library in New York City and Union Carbide in Tarrytown, NY; and taught library science courses (sci-tech reference materials) for the University of Rhode Island and SUNY Albany. She is also a professional indexer, and has been the editor of three newsletters.

Besides numerous articles, the majority for a column in the New England Association of Chemistry Teachers' semiannual journal *NEACT Journal*, she has written two books in addition to the 1982 version of the *Guidelines*.

Author's Acknowledgments

This work would not have seen the light of day without the encouragement of many people: my husband, Phillip Douville; my entire family; my editor at *NEACT Journal*, Judith Kelley, who suggested a book; and all the staff at *Choice*, who are the best co-workers ever.

Special thanks go to *Choice* Editor and Publisher Irving Rockwood, and *Choice* Managing Editor Fran Graf, who helped see this project to fruition. And, last but not least, to the reviewers for this list, mentioned previously: Thank you all very much for your hard work.

Extra special thanks are due Lisa Gross, *Choice* Production Manager, who put up with all the changes that were made at the last minute, and who is much appreciated by the author. And finally, thanks to David Durgin, who did the final copyediting, for his diligence and care.

To all my friends for being my friends, thank you.

Introduction

This work organizes chemical literature into ten areas, beginning with a brief standalone "guide" to the most important basic resources, then some general science describing resources that have a bearing on chemistry. The next chapters treat (in this order) analytical, physical, organic, inorganic, environmental, industrial, polymer, and biological chemistry. Each chapter is ordered into subsections reflecting the particular types of resources appropriate for each type of chemistry. Every effort was made to situate materials in appropriate places, based on their use and apparent main thrust. The final chapter presents some fairly important Internet sites, but not the entire universe of such sites. Sites come and go, and as these seem fairly stable they are offered to the reader for what they can do to enhance the chemical literature searching experience.

The material presented here had its origin in twice-yearly columns that appeared in *The NEACT Journal* published by the New England Association of Chemistry Teachers; the columns were intended to help chemistry teachers and professors cope with the burgeoning chemical literature, and to help students at all levels to find the correct works to answer their chemical questions. These columns have been gathered, information has been updated, and new material added as appropriate (new editions, replacements, completely new areas of study). Internet sources were included in the original columns; this material will be represented to some degree here.

Chapter 1

Basic Chemistry

Suppose you need the density of benzene. Where will you look? You'll find it in the "Rubber Handbook." But what if you want the density at some temperature not given in that handbook? Now the real detective work begins, and the investigation expands to include some of the more esoteric references such as the "Landolt-Bornstein" tables. How about a quick definition of carbon dating? Look in *Hawley's Condensed Chemical Dictionary*. What's "unslaked lime"? Another candidate for *Hawley's*. "Well," you say, "what's so hard about that? Everybody knows where to look for stuff like that." However, many do not know just where to find ready answers to many chemical questions. For example, what would you do if you needed to know all about diethylsilane, not just what it is, but also molecular weight, percent composition, molecular formula, preparation, physical and chemical properties, and some basic reactions? If you don't need it in vast detail, how about looking in the *Merck Index*? (Would you have known? Is it where you would look first?)

Time is often precious, and long hours in the library are not always possible for busy professionals. Your library (personal or institutional) may not have many reference works, and a trip to another library may be out of the question. These chapters form a series (by no means exhaustive) on chemical information sources, listing by category the large number of reference works in chemistry and allied fields. In these chapters you may find just the place to look for answers to questions, or to use substitutes if your chosen source is not readily available. Literature cited is current to the year 2003, with some exceptions.

Be careful! Libraries may classify these works differently, or place them in different locations; even though these listings contain Library of Congress and Dewey Decimal Classification numbers, the actual books may be located in other places, subject to the wishes of the library where they are shelved. Although searches through library stacks and reference departments can serendipitously yield great treasures, you should look up the locations of all desired reference works in the library's catalog before you set out on your investigation.

Enjoy your journey through the chemical literature!

Library Organization and Types of Literature Sources

Libraries are generally organized using two basic types of hierarchical classification schemes, the Dewey Decimal System (DDS), and the Library of Congress Classification Scheme (LC).

The DDS organizes chemistry, found in 500 (science), as follows:

540 — general
541 — physical
542 — experimental
543 — analytical
544 — qualitative analysis
545 — quantitative analysis
546 — inorganic
547 — organic
548 — crystallography
549 — mineralogy

Chemical engineering, a related field, is organized as follows:

660 — chemical engineering
661 — industrial chemicals
662 — explosives
663 — beverages
664 — food technology
665 — oils, fats, waxes, and gases
666 — ceramic and allied industries
667 — cleaning and dyeing
668 — other organic products
669 — metallurgy

The Library of Congress (LC) System has a similar breakdown for chemistry, part of Q (science):

QD — chemistry
QD 1-69 — general
QD 71-145 — analytical
QD 151-199 — inorganic
QD 241-449 — organic
QD 453-655 — physical and theoretical
QD 901-999 — crystallography

T, technology, is organized thus:

T — technology - general
TA — engineering - general
TC — hydraulic engineering

TD — sanitary and municipal engineering
TE — roads and pavements
TF — railroad engineering and operation
TG — bridges and roofs
TH — building construction
TJ — mechanical engineering and machinery
TK — electrical engineering and industries
TL — motor vehicles; cycles; aeronautics
TN — mineral industries; mining and metallurgy
TP — chemical technology
TR — photography
TS — manufactures
TT — trades
TX — domestic science (including cookery)

Chemical reference sources are characterized by their division into guides to the literature, book selection media, biographies, directories, meeting directories and calendars, as well as handbooks and tables of data, encyclopedias (both general and specific), technical and foreign language dictionaries, formularies, bibliographies, and safety literature. This material is often described as primary (journal articles, reports, patents), secondary (abstracting and indexing products), and tertiary (encyclopedias, directories, dictionaries), depending on the degree of reworking of the original material.

Journals, while not discussed here in great depth, are mentioned in a section following chapter 11.

Chemical information sources may also be divided up further using the major divisions of chemistry: analytical, organic, physical, inorganic, and so forth.

This first excursion into the chemical literature will begin with general reference literature specific to chemistry, followed by a section of general science literature references that are especially useful to chemists.

GUIDES TO THE LITERATURE

1.001

Allen, Barbara C., and Brian Livesey. *How to use Biological Abstracts, Chemical Abstracts, and Index Chemicus*. 2nd ed. Aldershot, Hampshire, England; Brookfield, VT: Gower, 1994. 103 p. 0566075563 pbk QD9 025.06 94-12968

Designed to assist medical personnel as well as biologists and chemists with searching techniques. Arrangement of each of the services is explained.

1.002

American Chemical Society. Division of Chemical Literature. *Searching the chemical literature*. Rev. and enl. ed. Washington, DC: ACS, 1961. 326 p. (Advances in chemistry series, 30). QD1 016 61-11330

Individually authored articles treat searching generally; patent searching; indexes and abstracts; *Chemical Abstracts*; nomenclature; the older chemical literature; searching for theses, dissertations, and unpublished information; house organs; medical literature; German and Soviet chemical literature; US government documents; and special libraries that have concentrations in chemical literature. Much of the information is very old but is good when information on older literature is needed.

1.003

Antony, Arthur. *Guide to basic information sources in chemistry*. Jeffrey Norton Publishers, Inc.; distributed by Wiley [1979]. 219 p. 0470265876 QD8.5 540 79-330

Treats guides to the literature, *Chemical Abstracts*, other abstracts and indexes, computerized searching, periodicals and periodical lists, primary publications (other than journals), bibliographies, nomenclature, dictionaries, encyclopedias, language dictionaries, general compilations of data, Beilstein, Gmelin, guides to techniques, safety manuals and guides, style guides, biographies and directories, company directories, nonprint materials, monographs, texts, treatises, and chemical information searching strategy.

1.004

Information sources in chemistry, ed. by Robert T. Bottle and J.F.B. Rowland. 4th rev. Bowker-Saur, 1993. 1857390164 QD8 540 92-17946

Discusses general sources, pure chemistry, and industrial chemistry. Separately authored chapters treat information and libraries, primary literature, abstracts and indexes, books, reviews, encyclopedias, online searching, chemical structure handling by computer, physicochemical data, inorganic and nuclear chemistry, Beilstein and other organic chemical sources, patents, commercial information, health and safety materials, pharmaceuticals, agrochemical and food chemistry, and national and international government information.

1.005

Lees, Nigel. *How to find information, chemistry: a guide* to searching in published sources. London: British Library, Science Reference and Information Service, 1995. 30 p. 0712308067 pbk QD8.5 016.54 95-22119

Includes brief information on chemical databases, information services, and other science reference materials.

1.006

Maizell, Robert E. *How to find chemical information: a guide for practicing chemists, educators, and students.* 3rd ed. NY: Wiley, 1998. 515 p. 0471125792 QD8.5 540 97-29120

Follows the same effective organization as the last two editions; 18 sections treat current awareness and document delivery; Chemical Abstracts Service (CAS); other abstracting and indexing services; US government technical information sources; various electronic media, including the Internet; critical literature reviews; major reference works; journals and patents; physical property data; health and safety information; marketing and business information; and analytical chemistry. New are appendixes with major chemical information awards, online databases for chemists and chemical engineers, and CAS databases available in the US. A new section, "Internet Issues and Tools," discusses pros and cons of use, caveats, and some sites of special interest. Also new is the section on CD-ROM "portable" databases. Of much importance is the expanded discussion of Beilstein, now available in a number of electronic versions. Patents, a very important source of chemical information, have been brought up to date, with material reflecting recent legal changes and new methods of searching online, including Internet resources. A new section on journals is very welcome, with publications of the American Chemical Society, the Royal Society of Chemistry (UK), and profiles of other organizations and publishers. Environmental, safety, and health materials have been extensively updated, with numerous electronic materials added and defunct services weeded out.

1.007

Mellon, Melvin Guy. *Chemical publications, their nature and use.* 5th ed. NY: McGraw-Hill, 1982. 419 p. 0070415145 QD8.5 540 81-20947

Part I: Publications, kinds and nature; general literature. Primary: periodicals, technical reports, patents, miscellaneous. Secondary: abstracts, indexes, reviews; *Chemical Abstracts*, bibliographies; tables, dictionaries,

encyclopedias, formularies; treatises; monographs; textbooks; nonprint materials. Tertiary: guides and directories. Part II: Publications, storage and use; libraries; manual and computerized searching; problem sets for students.

1.008

Ridley, Damon D. *Information retrieval: SciFinder and SciFinder Scholar.* Chichester: Wiley, 2002. 235 p. 0470843500; 0470843519 pbk Z699.5.S3 025.065

This book-length guide to the art and science of searching *Chemical Abstracts* begins with a concise introduction to the secondary literature and to indexing. Both *Chemical Abstracts* and *Medline* indexing conventions are explained, because SciFinder can be used for either. Chapters for each searching mode include "Research Topic," "Chemical Substance," and "Chemical Reaction." Each strategy is illustrated with screen prints that quickly familiarize users with what they will encounter.

1.009

Ridley, Damon D. *Online searching: a scientist's perspective: a guide for the chemical and life sciences.* Chichester; NY: Wiley, 1996. 344 p. 0471965200; 0471965219 pbk Z699.5.S3 025.06 95-54160

This was the first complete guide to searching STN, the most popular online source for scientists and the only one that contains *Chemical Abstracts* with complete records. Keywords, basic commands, and search terms are explained, as are contents and organization of some of the most important databases. Stepwise instructions for locating patent information and for doing chemical substance and structure searches.

1.010

Wiggins, Gary. *Chemical information sources.* NY: McGraw-Hill, 1991. 352 p. 2 disks. (McGraw-Hill series in advanced chemistry) 0079099394 QD8.5 540 90-61663

Describes online and interactive chemical databases; citation, subject, and patent searching; and online information sources other than abstracting and indexing services, such as dictionaries and formula handbooks. Covers structure searching, made possible (and easier) by online techniques, Other chapters describe traditional chemical information sources (properties, processes, chemical reactions, word meanings and translations, etc.) as well as toxicology, current awareness, and document delivery services. Offers comments on the personal library of a scientist, science writing aids, trade literature, standards information, and chemical industry information. Two disks contain the Chemistry Reference Sources Database (CRSD), with more than 2,150 records including references from the Indiana University Chemistry Library and the chemistry libraries at Purdue, University of Illinois, and University of Michigan.

1.011

Wolman, Yecheskel. *Chemical information: a practical guide to utilization.* 2nd rev. and enl. ed. NY: Wiley, 1988. 291 p. 0471917044 QD8.5 540.72 87-23119

Introduction; the scientific journal; the library: organization, locating material, online searching, sources outside the library, translations. Current awareness: meetings, research in progress, reviews, books, computerized services. How to conduct searches. Numerical data; chemical concepts; bibliographic databases; searching chemical reactions; structure/substructure searching. Environmental, patent, chemical marketing literature; people and organizations; expert systems.

ABSTRACTING AND INDEXING SERVICES

1.012

American Chemical Society. Chemical Abstracts Service. *Chemical Abstracts Service source index (CASSI).* 1970- . [Easton, PA]: ACS, 1971- . Quarterly. ISSN 0146-8065 Z5523x 80-13190

Cumulative edition covers 1907-99 in three volumes, with about 148,000 entries. Includes source material cited in Beilstein and *Chemisches Zentralblatt* back to 1830. Helps to locate libraries that have about 80,000 scientific and technical serial, periodical, or nonserial source publications. Valuable for the journal abbreviations included. Contains holdings of about 350 major resource libraries worldwide.

1.013

Chemical abstracts. American Chemical Society, 1907- . Columbus, Ohio, 1907- . v.1- . Weekly. ISSN 0009-2258 QD1 540.05 09-4698

"The Key to the World's Chemical Literature." Print, microform, and electronic formats. Volume indexes of authors, general subjects, chemical substances, and formulas.

Offers summary abstracts of more than 170,000 references annually, from all major languages worldwide. Covers journals, chemical patents, technical reports, dissertations, conference proceedings, and books; offers structure images; volume indexes every six months. Major subject sections: biochemistry; organic chemistry; macromolecular chemistry; applied chemistry and chemical engineering; physical, inorganic, and analytical chemistry. Author, patent, formula, general subject, chemical substance, and ring system indexes in cumulative volumes. *Chemical Abstracts,* authoritative *Registry Handbook— Common Names* on microfiche lists more than 720,000 compounds and more than a million names.

Chemical abstracts student edition is available through OCLC First Search at <http://www.oclc.org/firstsearch/>; it is suitable for smaller colleges and other institutions as a less-expensive alternative to the full *Chemical Abstracts.*

1.014

Chemisches zentralblatt. Jahrg. 1-140 Berlin, Verlag Chemie, 1830-1969. v.1-140. Weekly. QD1 540.5

Up to 1919, limited to articles on pure chemistry only, of German chemical research. From 1919, covered worldwide literature for pure and applied chemistry. Considered to be the most important abstracting journal because of the length of time covered. Abstracts were classified as follows: A: General; physical; inorganic chemistry (9 subdivisions); B: General and theoretical organic chemistry; C: Preparative organic chemistry; naturally occurring substances (8 subdivisions); D: Macromolecular chemistry; E: Biological chemistry; physiology; medicine (6 subdivisions); F: Pharmaceutical chemistry; disinfections; G: Analysis; laboratories; H: Applied chemistry (24 subdivisions). From 1830 to 1848, titled *Pharmaceutisches Zentralblatt;* 1849-1855, *Chemisch-Pharmaceutisches Centralblatt;* 1856-1896, *Chemisches Centralblatt;* and finally, 1897-1969, *Chemisches Zentralblatt.*

ENCYCLOPEDIAS

1.015

Concise encyclopedia chemistry, tr. and rev. by Mary Eagleson. English language ed. Berlin; NY: Walter de Gruyter, 1994. 1,201 p. 0899254578; 3110114518 QD4 540 93-36813

A translation and revision of the 5th edition of the German language *ABC Chemie* (1993). Most of the 12,000 entries are about 15 lines, although some (e.g., nomenclature) are much longer, running up to 20 pages. Definitions and explanations are usually succinctly written and illustrated when appropriate. Approximately one-third of the entries are for specific compounds and usually contain information on sources, properties, abundance, reactions, and applications as well as structural diagrams. No CAS Registry Numbers nor entries for individual scientists; lists Nobel Prize winners in chemistry and related fields. SI units and IUPAC rules are used throughout.

1.016
The encyclopedia of the chemical elements, ed. by Clifford A. Hampel. NY: Reinhold Book Corp., 1968. 849 p. QD466 546.11 68-29938

Bibliographies included. Signed short paragraphs/articles in alphabetic order. Articles treat the 103 elements: history, discovery, occurrence, nuclear properties, preparation and isolation, physical and chemical properties, compounds, detection and determination, raw materials, physiological behavior, isotopes. Illustrations.

1.017
[Kirk-Othmer] Encyclopedia of chemical technology. 2nd ed. compl. rev. Herman Mark, John J. McKetta Jr., and Donald F. Othmer, ed. board. NY: Interscience, 1963-70. 22 v.

[Kirk-Othmer] Encyclopedia of chemical technology. 3rd ed. Martin Grayson and David Eckroth, executive eds. NY: Wiley, 1978-84. 31 v. 0471020370 TP9 660

[Kirk-Othmer] Encyclopedia of chemical technology. 4th ed. Jacqueline I. Kroschwitz, executive ed. NY: Wiley, 1991-98. 25 v.; suppl.; index to v. 1-25 and suppl. 047152669x TP9 660 91-16789

These three encyclopedia sets are extremely valuable for getting the entire picture concerning a particular chemical subject. Older editions of the encyclopedia are retained by libraries because material is not repeated in subsequent editions if there is nothing newer on the subject; the more recent editions will refer to material in the previous editions. The 4th edition provides broad but comprehensive coverage of applied chemistry and chemical technology. The text retains its familiar, readily accessible format, but all articles were rewritten; most authors are from major industrial laboratories, reflecting a shift in approach. About half of the encyclopedia treats specific chemical compounds or groups of compounds. Much more emphasis is now placed on health hazards and safety precautions in articles discussing individual compounds and classes of compounds than in the previous edition. Other areas receiving expanded coverage include biotechnology, chemical analysis, computer applications, economic considerations, energy, and the environment. An absolutely indispensable reference resource.

1.018
Kirk-Othmer online. NY: John Wiley & Sons, Inc.; Available via subscription from vendors and Wiley.

Covers 3rd ed. and forward of *Kirk-Othmer Encyclopedia of Chemical Technology* (1984-); designed to stay current and will be revised and updated as the industry develops. The full text of the print version is available in this file.

1.019
Kirk-Othmer concise encyclopedia of chemical technology. 4th ed. NY: Wiley, 1999. 4th ed. 2,196 p. Exec. editor, J. Kroschwitz. (A Wiley-Interscience publication) 0471296988 TP9 660 98-50686

Abridged version of the 26-volume *Kirk-Othmer Encyclopedia of Chemical Technology*, 4th ed., 1991-98. Follows the same format as the full version and includes shortened versions of the original text, graphs, illustrations, and references.

1.020
Macmillan encyclopedia of chemistry; Joseph J. Lagowski, ed. in chief. NY: Simon & Schuster Macmillan, 1997. 4 v. 0028972252 (set); 002897221x (v. 1); 0028972228 (v. 2); 0028972236 (v. 3); 0028972244 (v. 4) QD4 540 97-1824

Some 730 short and long articles, alphabetically arranged, that treat basic terms and concepts within the major areas of matter, principles and laws of chemistry, the main divisions of chemistry, aqueous chemistry, biochemistry, industrial applications, and teaching. Inside v. 1, an outline of contents. More than 1500 illustrations of chemical structures, equations, and principles. Includes 100 biographical entries on chemists who made seminal contributions. Articles contain cross- and *see* references and end with up-to-date bibliographies. More than 1,500 illustrations of chemical structures, equations, and principles. Common abbreviations and symbols, a list of journal abbreviations, and a periodic table appear at the beginning of each volume, and a complete index in the fourth. Article contributors and their affiliations are listed.

1.021

Van Nostrand Reinhold encyclopedia of chemistry;
ed. in chief, Douglas M. Considine; managing ed., Glenn D.
Considine. NY: Van Nostrand Reinhold, 1984. 4th ed.
1,082 p. 0442225725 QD5 540 83-23336

Rev. ed. of *The Encyclopedia of Chemistry*, 3rd ed., 1973.
This edition contains 85% revised material, 500 more entries,
and includes about 1,300 topics. The preface lists 11 subjects
that received particular emphasis. Detailed coverage on
advanced processes, strategic raw materials, chemistry of
metals, energy sources and conversion, wastes and pollution,
analytical instrumentation, food chemicals, structure of
matter, and new materials. Somewhat dated, but good for a
beginning search.

DICTIONARIES

1.022

Bennett, Harry. *Concise chemical and technical
dictionary*. 4th enl. ed. NY: Chemical Pub. Co., 1986. 1,271
p. 0820602043 QD5 540.3 87-150905

1st ed., 1947. More than 85,000 brief definitions of
scientific terms, chemicals, trademarked products, and drugs.

1.023

A Concise dictionary of chemistry. Oxford; NY: Oxford
Univ. Press, 1990. New ed. 314 p. (Oxford reference)
0192861107 QD5 540 90-38630

This handy volume is derived from the earlier *Concise
Science Dictionary*. The scope includes all areas of
chemistry and much biochemistry. There are more than
3,000 entries, many of which reflect recent advances in
techniques, concepts, and materials. Diagrams; table of SI
measurements; appendixes of fundamental constants; solar
system; and periodic table of the elements.

1.024

A dictionary of chemistry, ed. by John Daintith. 4th ed.
Oxford; NY: Oxford University Press, 2000. 586 p. (Oxford
paperback reference) 0192801015 pbk QD5 540 2001-
268453

Fully revised; contains more than 4,000 entries. Includes
physical and biological chemistry terms, and treats laboratory
techniques, chemical engineering, and environmental matters.
Table of elements with properties. Available online via
OxfordReference Online, <http://www.oxfordreference.com/>.

1.025

Hackh, Ingo Waldemar Dagobert. *Grant and Hackh's
chemical dictionary: American and British usage;
containing the words generally used in chemistry, and
many of the terms used in the related sciences of physics,
medicine, engineering, biology, pharmacy, astrophysics,
agriculture, mineralogy, etc., based on recent scientific
literature*. 5th ed., completely rev. and ed. by Roger Grant,
Claire Grant. NY: McGraw-Hill, 1987. 641 p. 0070240671
QD5 540.3 86-7496

Rev. ed. of *Chemical Dictionary*, 4th ed., 1969. Prefatory
section of explanatory notes. Material arranged in true
dictionary style, with definitions in few words or sentences.

1.026

Hampel, Clifford A., and Gessner G. Hawley.
Glossary of chemical terms. 2nd ed. NY: Van Nostrand
Reinhold, 1982. 306 p. 0442238711 QD5 540 81-11482

Short paragraphs to explain terms. Some drawings. About
2,000 terms used in chemistry and the process industries.
Brief biographical sketches. Cross-references.

1.027

Hawley's condensed chemical dictionary. 14th ed.,
rev. by Richard J. Lewis, Sr. NY: Wiley, 2002. 1,223 p.
0471387355 (print); 0471055328 (CD-ROM); 0471055336
(print/CD-ROM) QD5 540 2001-45614

Alphabetic listing (two columns per page) in dictionary
format. Definitions contain synonyms, CASRNs, molecular
formulas, properties, hazards, uses. List of abbreviations.
Appendixes. Web links to manufacturers and associations.
Updated information on production, usage, and regulatory
trends.

1.028

*Hawley's condensed chemical dictionary: the VNR
dictionary of environmental health and safety; Sax's
dangerous properties of industrial materials
[computer file]*. NY: Van Nostrand Reinhold, 1995. 1 CD-
ROM 0442020449 Q180

This electronic product includes three well-known
references: *Hawley's Condensed Chemical Dictionary*,
originated by Gessner Goodrich Hawley, 12th ed.; *VNR
Dictionary of Environmental Health and Safety*, ed. by Frank
S. Lisella; and *Sax's Dangerous Properties of Industrial
Materials*, originated by N. Irving Sax, 8th ed.

1.029

Howard, Philip H., and Michael Neal. *Dictionary of chemical names and synonyms*. Boca Raton, FL: Lewis Publishers, 1992. 1 v., various pagings (2,528 p.) 0873713966 TP9 660 92-9160

A one-volume dictionary of about 20,000 chemicals and 130,000 names, intended both to identify synonyms and to find CASRN. The principal part is a 1,400-page numerical list, giving for each registry number the molecular formula and alternate names, popular names, trade names, and code numbers. The second part, an alphabetical index of all these synonymous names, provides only CASRN. A 90-page index by molecular formula provides CASRN.

1.030

McGraw-Hill dictionary of chemistry, ed. in chief, Sybil P. Parker. NY: McGraw-Hill, 1997. 454 p. 0070524289 QD5 540 96-46184

Material in this volume has been extracted without change from the *McGraw-Hill Dictionary of Scientific and Technical Terms*, 5th edition, which is available online from Access Science <http://www.accessscience.com/>. Generous vocabulary: for chemistry, more than 8,000 terms (publisher's estimate). The illustrations that were an important feature of the basic dictionary have been dropped. Appendixes of conversion factors and other tables of importance.

1.031

The Penguin dictionary of chemistry, ed. by David W.A. Sharp. 3rd ed. London: Penguin, 2003. 480 p. (Penguin reference) 0140514457 540.3

Explains chemical terms from all branches of chemistry; illustrated throughout.

1.032

Wohlauer, Gabriele E. M., and H.D. Gholston. *German chemical abbreviations*. NY: Special Libraries Association, 1965. 63 p. QD7

Technical and some nontechnical abbreviations; some Latin entries; information presented in parallel columns, with abbreviation in German or Latin and the corresponding word or phrase in full, with English equivalent. About 2,500 entries.

FOREIGN LANGUAGE DICTIONARIES

1.033

Callaham, Ludmilla Ignatiev, Patricia E. Newman, and John R. Callaham. *Callaham's Russian-English chemical and polytechnical dictionary*. 4th ed. NY: Wiley, 1996. 814 p. 0471611395 QD5 603 94-37599

Supplies more than 120,000 Russian terms in the physical chemical, life science, and engineering areas, and some 5,000 most frequently used nontechnical terms. Comprehensive translations of Russian verbs and meaning variations for different contexts. Information on how Russian prefixes, suffixes, and roots are combined to form new terms.

1.034

Dictionary of chemical terminology, in five languages: English, German, French, Polish, Russian. Ed. by Dobromila Kryt. Amsterdam; NY: Elsevier Scientific Publ. Co., 1980. 600 p. 0444997881 QD5 540 79-20852

Based on the earlier Polish dictionary *Slownik terminologii chemicznej polsko-niemiecko-angielsko-francnsko-rosyjski*, published in 1974. Coverage emphasizes terminology from chemistry, but includes that from related disciplines, gathered from international publications. Excludes chemical compound nomenclature and terms from chemical engineering. The 3,805 entries are defined in English followed by the other language equivalents. Cross-references; some illustrations. Alphabetical indexes in all the languages.

1.035

Dictionary of common names = Trivialnamen-Handbuch, ed. by FIZ Chemie, Berlin. Weinheim; NY: VCH, 1993. 3 v., 2,464 p. 3527290206 (Weinheim, set); 1560817208 (NY, set) QD291 547 93-232093

This handbook pairs chemical names with the compounds they represent. The first two volumes list about 22,000 chemicals in a single alphabetic sequence according to their most common German trivial name (usually very similar to the English equivalent). For each name are given the usual German name, English translation, very clear stereochemical drawing, molecular formula, CASRN, and recent literature references. No full standard chemical names are included. V. 3 has brief German and English indexes of alternative trivial names.

1.036

Kaplan, Steven M. *Wiley's English-Spanish, Spanish-English chemistry dictionary = Diccionario de química inglés-español, español-inglés Wiley.* Steven M. Kaplan, lexicographer; editorial advisors, Rolf Altschul, Sylvia Márquez-Pirazzi. NY: Wiley, 1998. 530 p. (A Wiley-Interscience publication) 0471192880; 0471249238 pbk QD5 540 97-34936

The English section contains more than 40,000 words and phrases followed by Spanish equivalents; the Spanish portion offers the reverse. Includes all areas of chemistry, including biochemistry.

1.037

Langenscheidt Routledge German dictionary of chemistry and chemical technology = Wörterbuch Chemie und chemische Technik Englisch, ed. by Technische Universität Dresden. 5th ed. London; NY: Routledge, 1997- . 2 v. 0415171288 (v. 1); 0415173361 (v. 2) QD5 540 97-33480

V. 1: German English; v. 2: English-German. Contains about 3,000 new headwords in organic chemistry, physical chemistry, and biochemistry, as well as chemical technology, making a total of about 50,000 words. Definitions consist of a single word or phrase, with some parenthetical explanations of context. Gender is provided for German nouns, and both British and American spellings are included.

1.038

Patterson, Austin McDowell. *French-English dictionary for chemists.* 2nd ed. NY: Wiley, [1954]. 476 p. QD5 540.3 54-6661

Abbreviations list. Introduction discusses conventions, word forms, nomenclature. French-English only; no reverse dictionary.

1.039

Patterson, Austin McDowell. *Patterson's German-English dictionary for chemists.* 4th ed. Revised by James C. Cox; edited by George E. Condoyannis. NY: Wiley, 1992. 890 p. 0471669911 QD5 540 90-45365

This new edition lists some 65,000 terms. Some terms labeled "Old Chem" in the 3rd ed. have been dropped to make room for newer terms, so that edition should not be discarded. New features are discussed in a lengthy and essential essay.

1.040

Wenske, Gerhard. *Wörterbuch Chemie: Deutsch/Englisch = Dictionary of chemistry: German/English.* Weinheim; NY: VCH, 1994. 2,041 p. (Parat, ISSN 0930-6862) 3527264299 (Weinheim); 089573527x (NY) QD5 540 95-100206

English translations for approximately 150,000 German words and phrases, for all chemical and related sciences, including biochemistry and chemical engineering. Provides idiomatic modern equivalents and specifies appropriate meanings for related words or for words used in special senses in specific fields. All noun-adjective combinations are listed twice, once following the listing for the main noun, and once under the adjective. A full selection of names of specific chemicals is included, even when essentially identical in the two languages. Chemical formulas and IUPAC or other systematic designations are specified for about half of the chemicals. Exceptionally readable, with much clearer presentation and larger type than is customary in such works.

HANDBOOKS AND TABLES OF DATA

1.041

Aldrich Chemical Company. *Catalog handbook of fine chemicals.* Milwaukee, WI: Aldrich Chemical Co., 1988-89- . Biennial. TP202 661.0029 89-38391

(Also issued 1990-91- on CD-ROM as: Aldrich Chemical Company. *Aldrichem Data Search*, and on diskette as: *Aldrich Catalog on Diskette.*) "Aldrich" is an example of a catalog of commercially available chemicals that includes properties data and references to major sources of information. This book is usually kept close at hand in many laboratories. Chemicals, including reagents, are arranged alphabetically; there are many cross-references, including from acronyms commonly used for reagents, solvents, etc. Includes a molecular formula index, if a check for the substance by name is not productive. Provides: property data such as melting and boiling points; density; safety and toxic information such as RTECS no., terms that describe toxicity; and references to safety compendia; molecular formula; CASRNs; and structures for more complicated substances. Provides references to such sources as *Merck Index*, IR and NMR spectra, Beilstein, Fieser's *Reagents for Organic Synthesis*, and toxicity compilations.

1.042

Composite index for CRC handbooks [computer file], 3rd ed., software version 1.1. Boca Raton, FL: CRC Press, 1992. CD-ROM; two 3.5" disks; user's guide. 0849302900 QD65 540 93-790484

Provides a composite index to 368 handbooks on chemistry and mathematics published by CRC Press. Includes 1991 supplement.

1.043

Coyne, Gary S. *The laboratory companion: a practical guide to materials, equipment, and technique.* NY: Wiley-Interscience, 1997. rev. ed. 527 p. (A Wiley-Interscience publication) 0471184225 QD53 542 97-16689

This book is an updated version of Coyne's *The Laboratory Handbook of Materials, Equipment, and Technique*, 1992. His expertise and experience in building and repairing research apparatus has provided unusual insights clearly and succinctly articulated in the new version. The book is an excellent guide to laboratory materials, research equipment, and supplies, and to measurement and operational techniques in the laboratory. A valuable part is the emphasis on laboratory safety. Explains the rationale for laboratory procedures and provides historical background and helpful guidelines and procedures for the techniques discussed.

1.044

CRC handbook of chemistry and physics: a ready-reference book of chemical and physical data; ed. in chief, David R. Lide. 84th ed. Boca Raton, FL: CRC Press, 2003. Annual edition. Sections separately paged. 0849304849 QD65 540.212 13-11056

Arrangement: mathematical tables; elements and inorganic compounds; organic compounds; general chemical information; general physical constants; miscellaneous (conversion tables and factors); the section "Physical Constants of Organic Compounds" is now arranged alphabetically by substance name, and structure diagrams are integrated with data.

1.045

CRC handbook of chemistry and physics [computer file]; CRCnetBASE; ed. in chief, David R. Lide. Boca Raton, FL: Chapman & Hall/CRCnetBASE, 1999- . Annual. 1 CD-ROM (Part of the Chapman and Hall/CRCnetBASE series) ISSN 1098-4178 QD65 530 98-2006

Contains all the content of the print version, plus additional information and searching capabilities including cross-table searching, index and synonym browsing, unit conversions, structure viewing, complete text/word searching; extensive Boolean and proximity searching. The mathematical tables appendix is omitted. Information can be accessed in browse or search modes. Also available via the Internet.

1.046

Faust, Rüdiger, Günter Knaus, and Ulrich Siemeling. *World records in chemistry;* Hans-Jürgen Quadbeck-Seeger, ed.; transl. by William E. Russey; [with a foreword by Ronald C.D. Breslow]. 1st English ed. Weinheim; NY: Wiley-VCH, 1999. 361 p. 3527295747 QD37 540 99-230079

The main body of the book is divided into 23 sections with titles ranging from "Atoms and Molecules," "Chemical Industry," "Literature," "Mistakes," "Molecular Form," "Pharmaceuticals," "Poisons," "Sensors," "Syntheses," and so on. Easy-to-read entries are clearly and concisely written, and there are numerous tables and figures. Useful name and subject indexes.

1.047

Gardner's chemical synonyms and trade names, ed. by G.W.A. Milne. 11th ed. Brookfield, VT: Ashgate, 1999. 1,418 p. 0566081903 TP9 660 98-51143

Formerly edited by Michael and Irene Ash. Some 35,000 entries list physical properties, chemical composition, applications, suppliers, and CASRN, European Inventory of Existing Commercial Chemicals (EINECS), and *Merck Index* entry numbers where available. Chemicals are arranged alphabetically by trade name, with a section containing firm names. Index of manufacturers. Also available in digital version as *Gardner's Digital Handbook, Version 2.0*, on CD-ROM.

1.048

Gordon, Arnold J., and Richard A. Ford. *The chemist's companion: a handbook of practical data, techniques, and references.* NY: Wiley, [1972]. 537 p. 0471315907 QD65 542 72-6660

A somewhat dated handbook, but included here for basic information. More emphasis on techniques than other data compilations. Sections on properties of molecular systems; properties of atoms and bonds; kinetics; energetics; spectroscopy; photochemistry; chromatography;

experimental techniques; mathematical and numerical information.

1.049

Lange, Norbert Adolph. *Lange's handbook of chemistry,* ed. by John A. Dean. 15th ed. NY: McGraw-Hill, 1999. About 1,100 p. in various pagings. 0070163847 QD65 540 84-643191 (for 14th ed.)

Eleven sections: mathematical information; general information and conversion tables; atomic and molecular structure; inorganic chemistry; analytical chemistry; electrochemistry; organic chemistry; thermodynamic properties; physical properties; and miscellaneous.

1.050

Merck index; an encyclopedia of chemicals, drugs, and biologicals. 13th ed. Maryadele J. O'Neil, senior ed.; Ann Smith, senior assoc. ed.; Patricia E. Heckelman, assoc. ed. Whitehouse Station, NJ: Merck, 2001. 1 v., 2,564 p. in various pagings. 0911910131 RS51 615.103 89-60001

Information on human and veterinary drugs, biologicals and natural products, agricultural chemicals, industrial and laboratory chemicals, and environmentally significant materials, described in more than 10,000 monographs (concise description of a single material or small group of closely related materials). Chemicals (numbered) arranged alphabetically; CASRN, therapeutic category, and biological activity index; formula index; cross-index of names; titles of deleted "monographs" from previous editions; appendix (addenda).

1.051

The Merck index [computer file] (Merck index on CD-ROM) Ver. 13.0 Cambridge, MA: CambridgeSoft, 2001. 1 CD-ROM RM51

Electronic equivalent of *Merck Index*, 13th ed. Also available online via the Internet from CambridgeSoft.

1.052

Ockerman, Herbert W. *Illustrated chemistry laboratory terminology*. Boca Raton, FL: CRC, 1991. 211 p. 0894301521 QD51 542 91-8591

Lists hundreds of items encountered in the typical chemistry laboratory. Intended for students whose first language is not English, but the illustrations have English captions. Thorough table of contents in English, Chinese, French, German, Polish, Spanish, and Turkish. Items are named in British English terms or spelling along with the American version.

1.053

Shugar, Gershon J., and Jack T. Ballinger. *The chemical technicians' ready reference handbook*. 4th ed. Consulting ed., Linda M. Dawkins. NY: McGraw-Hill, 1996. 972 p. 0070571864 QD61 542 95-52614

Almost half new material over the previous edition. The 38 chapters provide chemical and laboratory technicians with information on performance of duties accurately, safely, and professionally. Presents latest information on federal requirements, lab safety, and stepwise operational procedures for current chemical technologies.

1.054

Shugar, Gershon J., and John A. Dean. *The chemist's ready reference handbook*. Consulting ed., Ronald A. Shugar, et al. NY: McGraw-Hill, 1990. 640 p. 0070571783 QD65 543 89-8166

Collects data, procedures, precautions, and troubleshooting information for laboratory chemical work. Treats theory briefly but offers information on real-world questions and problems encountered during laboratory work. Describes laboratory instruments and use; details analytical procedures; offers precautions and safety measures; and gives checklists and troubleshooting hints.

DIRECTORIES

1.055

American Chemical Society, Committee on Professional Training. *Directory of graduate research*. Washington, DC: The Society, 1953- . Biennial. ISSN 0193-5011 QD40 016.54 79-3432

Includes data on faculties, publications, and doctoral and master's theses in departments or divisions of chemistry, chemical engineering, biochemistry, pharmaceutical/medicinal chemistry, clinical chemistry, and polymer science at universities in the U.S. and Canada. Institutions are listed alphabetically; for each department is presented degrees offered; fields of specialization; faculty with rank, personal data, telephone numbers, and research publications; and statistical information. A CD-ROM, Version 1.0, became available in 1993 (ACS Directories On Disc; a Web version can be found at the ACS site, via subscription, as *DGR Web*).

1.056

Directory of chemistry software, ed. by Wendy Warr, Peter Willett, and Geoff Downs. Oxford: Cherwell Scientific; Washington, DC: American Chemical Society, 1992. 204 p. Annual. 0951823604 QD39.3.C6 92-19715

This directory guides the user to software that manipulates structural information. Software categories include structure drawing, database management systems, expert systems, 3-D molecular modeling, quantum mechanics, utility packages, terminal emulators for online searching, and communications and off-line query formulation. Main categories of hardware requirements for software in the directory are IBM PCS and compatibles, Apple Macintosh, workstations running Unix, DEC VAX, mainframes, and supercomputers. More than 170 software packages available internationally are listed alphabetically by program name with each package description on a single page. Product description provides company and distributor contact information, product features and capabilities, hardware and software requirements, level of support given, price, and available references to product reviews. List of publishers/distributors; program index of systems.

1.057

World databases in chemistry, ed. by C.J. Armstrong. London; NJ: Bowker-Saur, 1996. 1,200 p. (World databases series) 1857391012 QD8.3 025.06 98-196026

The main section is divided into nine classes; within each area, databases are listed alphabetically, and every unique database is described in a master record. Following this record are descriptions of different implementations of the database, e.g., CD-ROM, tapes, specific online vendors, etc.; chronological portions or other subsets of databases are listed after the master record. Databases relevant to more than one section are listed and described in each section. Descriptions (about one-half page) include comments by reviewers and provide generally reliable critical evaluations, fairly detailed content summaries, and information about prices, update frequency, search aids, and software. Database name index with more than 3,000 entries and a directory containing e-mail addresses and telephone, telex, and fax numbers.

STYLE GUIDE

1.058

The ACS style guide: a manual for authors and editors. 2nd. ed. Washington, DC: American Chemical Society, 1997. 460 p. 0841234612; 0841234620 pbk QD8.5 808 96-49413

A newer edition of *Handbook for Authors of Papers in American Chemical Society Publications*, 3rd ed., 1978. Discusses style for various journals, not limited to ACS publications (some other society publications follow ACS style).

BIOGRAPHIES

1.059

American chemists and chemical engineers. Ed. by Wyndham D. Miles. Washington: American Chemical Society; Guilford, CT: Gould Books, 1976-<1994 >. v. <1-2 >; v. 1, 544 p. 0841202788; 0964025507 (v. 2) QD21 540 76-192

V. 2 ed. by Wyndham D. Miles and Robert F. Gould. Contains biographies of important chemists and chemical engineers in the U.S. V. 1 records the lives and careers of 517 men and women, preeminent in chemistry over the past 300 years.

1.060

Barkan, Diana Kormos. *Walther Nernst and the transition to modern physical science*. Cambridge, UK; NY: Cambridge Univ. Press, 1999. 288 p. 052144456x QD22.N39 540 98-22028

Nernst (1864-1941) received the 1920 Nobel Prize in Chemistry, particularly for thermochemistry. As a physical chemist strongly rooted in physics, Nernst is remembered especially for the Nernst equation of electrochemistry, for the Nernst glower or globar, and for the Nernst heat theorem, which became the basis for the Third Law of Thermodynamics. Nernst obtained patents for his lamp or glower and profited financially from its industrial development. Emphasizes the importance of this lamp and related experimental work to the development of the heat theorem; discusses the science and personal tensions related to Nernst's somewhat delayed Nobel Prize, including the role of Svante Arrhenius.

1.061

Basolo, Fred. *From Coello to inorganic chemistry: a lifetime of reactions*. NY: Kluwer Academic/Plenum Publishers, 2002. 245 p. (Profiles in inorganic chemistry) 0306467747 QD22.B258 540 2002-22220

Basolo (Northwestern University) presents this excellent travelogue, a very personal account of a chemist who played a major role in the advancement of inorganic chemistry after World War II. Born in Coello, Illinois, the youngest son of poor Italian immigrant parents, he became president of the American Chemical Society and received almost all applicable domestic and international awards for excellence in chemical research and teaching.

1.062

Bowden, Mary Ellen. *Chemical achievers: the human face of the chemical sciences*. Philadelphia: Chemical Heritage Foundation, 1997. 180 p. 0941901130 pbk QD21 540 97-5508

Includes pictures and brief biographies of 80 chemists and related scientists, and some illustrations. The time span ranges from the 17th century (Robert Boyle) to the present (including scientists still living), with a majority representing the 20th century. Ten women are covered, including Lise Meitner and Rosalind Franklin. Chemistry and the chemical sciences are considered broadly, including traditional chemistry, biochemistry and molecular biology, nuclear chemistry and physics, chemical industry, chemical engineering, and the environment. Much of the value of this charming book lies in this broad view of the chemical sciences and the faces of the practitioners. Extensive bibliography.

1.063

Candid science: conversations with famous chemists; István Hargittai; ed. by Magdolna Hargittai. London: Imperial College Press; River Edge, NJ: Distributed by World Scientific Pub., 2000. 516 p. 1860941516; 1860942288 pbk QD21 540.922

Interviews with 36 famous chemists, including 18 Nobel laureates, conducted by Hargittai.

1.064

Djerassi, Carl. *The pill, pygmy chimps, and Degas' horse: The autobiography of Carl Djerassi*. NY: Basic Books, 1992 319 p. 0465057594 QD22 540 91-58542

Describes the early and middle years of Djerassi's life, as he came to the US from Europe, and became one of the

world's foremost organic chemists; later known for work on contraceptives as "Father of the Pill."

1.065

Chemistry, 1901-1921. Nobel lectures in chemistry, 1901-1921. Singapore; River Edge, NJ: World Scientific, 1999. 409 p. 9810234058 QD39 540

Chemistry, 1922-1941. Nobel lectures in chemistry, 1922-1941. Singapore; River Edge, NJ: World Scientific, 1999. 506 p. 9810234066 QD39 540

Chemistry, 1942-1962. Nobel lectures in chemistry, 1942-1962. Singapore; River Edge, NJ: World Scientific, 1999. 710 p. 9810234074 QD39 540

Chemistry, 1963-1970. Nobel lectures in chemistry, 1963-1970. Singapore; River Edge, NJ: World Scientific, 1999. 357 p. 9810234082 QD39 540

Chemistry, 1971-1980. Nobel lectures in chemistry, 1971-1980. Ed. in charge, Tore Frängsmyr; ed., Sture Forsaen. Singapore; River Edge, NJ: World Scientific, 1993. 447 p. 9810207867; 9810207875 pbk QD39 540 95-211946

Chemistry, 1981-1990. Nobel lectures in chemistry, 1981-1990. Ed. in charge, Tore Frängsmyr; ed., Bo. G. Malmstrom. Singapore; River Edge, NJ: World Scientific, 1992. 708 p. 9810207883; 9810207891 pbk QD39 540 96-142427

Chemistry, 1991-1995. Nobel lectures in chemistry, 1991-1995. Singapore; River Edge, NJ: World Scientific, 1997. 296 p. 9810226799; 9810226802 pbk QD39 540 97-178966

All these volumes contain Nobel lectures, including presentation speeches and laureates' biographies.

1.066

Farber, Eduard. *Great chemists*. NY: Interscience, 1961. 1,642 p. QD21 925 60-16809

Contains 114 biographies taken from literature as far back as possible; some extend back to Babylonian times.

1.067

Goertzel, Ted George, and Ben Goertzel. *Linus Pauling: a life in science and politics*, with the assistance of

Mildred Goertzel, Victor Goertzel, with original drawings by Gwen Goertzel. NY: Basic Books, 1995. 300 p. 0465006728 Q143.P25 540 95-9005

Research for this biography of Pauling (1901-94) was carried out by three generations of the Goertzel family over some 30 years. It is not an authorized biography, but Pauling was cooperative and reviewed some of the manuscript before his death. The authors, in the academic fields of sociology and cognitive science, discuss appreciatively and critically the major aspects of Pauling's life, career, accomplishments, and impact. Ample notes and references.

1.068
Jaffe, Bernard. *Crucibles: the lives and achievements of the great chemists.* NY: Simon and Schuster, 1930. 377 p. QD21 540 36-7247

Contains 16 biographies and portraits of chemists living from 1404 to 1881, from Bernard Trevisan (1406-90) up to Irving Langmuir (b. 1881). About 15 pages per chapter (1934 edition from Tudor Publ. Co., NY).

1.069
Jaffe, Bernard. *Crucibles: the story of chemistry from ancient alchemy to nuclear fission.* New rev. and updated 4th ed. NY: Simon and Schuster, 1948. 480 p. Reprinted: NY: Dover Publications, 1976. 368 p. 0486233421 QD21 540 75-38070

This book, and the previous entry, contain biographies of popular chemists.

1.070
Knight, David N. *Humphry Davy: Science & power.* Oxford, UK; Cambridge, MA: B. Blackwell, 1992. 218 p. (Blackwell science biographies) 0631168168 QD22 540 92-11582

Outstanding biography of the great 19th-century chemist Davy. Complete notes; select bibliography; full index.

1.071
Lewis, Edward S. *A biography of distinguished scientist Gilbert Newton Lewis.* Lewiston, NY: Edwin Mellen Press, 1998. 119 p. 0773482849 QD22.L57 540 98-25084

Useful, brief synopsis of the career of one of the most distinguished American chemists by his son. It includes too few reminiscences of Lewis the man but does include some 20 photographs of Lewis and his family plus seven pages of appreciations by well-known scientists. Lewis pioneered in the electronic theory of valency and contributed significantly to chemical thermodynamics, especially in the studies of free energies. He is especially remembered for his very generalized definition of acids and bases; the term "Lewis acid" is now in common use in the chemical world. Full bibliography of Lewis's publications.

1.072
McGrayne, Sharon Bertsch. *Prometheans in the lab: chemistry and the making of the modern world.* NY: McGraw-Hill, 2001. 243 p. 0071350071 TP18 660 2001-30671

A striking and readable collection of nine thumbnail biographies of heroic (and troubled) figures in the history of chemistry, from Revolutionary France to Superfund America. Chemists will be familiar with some of the biographees (Perkin, Carothers, Haber), but some (Rilleux, Midgely) are not even laboratory, let alone household, familiars. Yet all of them did work that is central to the creation of a world of clean air, safe food and water, cheap transport, and long lives.

1.073
Nobel laureates in chemistry, 1901-1992. Ed. by Laylin K. James. Washington, DC: American Chemical Society; Philadelphia: Chemical Heritage Foundation, 1993. 797 p. 0841224595; 0841226903 pbk QD21 540.92 93-17902

Arranged in chronological order, with dates, bibliographies. Footnoted text, with five to seven pages in each biography.

1.074
Olah, George A. *A life of magic chemistry: autobiographical reflections of a Nobel Prize winner.* NY: Wiley-Interscience, 2001. 277 p. 0471157430 QD22.O43 540 00-43638

Olah reminiscences on growing up in Hungary, early research and teaching; moving to North America; industrial experience; returning to academa; building the Loker Institute; Nobel Prize-winning research, and society and chemistry.

1.075
Poirier, Jean Pierre. *Lavoisier, chemist, biologist,*

economist; transl. from the French by Rebecca Balinski. Philadelphia, PA: University of Pennsylvania Press, 1996. 516 p. (The chemical sciences in society series) 0812233654; 0812216490 pbk QD22.L4 540 96-35738

This major biography of Lavoisier discusses the economic and political dimensions of his career. It details his services to the French Academy of Sciences and to the many governmental commissions and boards to which he was appointed during both the time of the monarchy and the turbulence of revolution. Some illustrations; full notes; bibliography; name and subject indexes.

1.076

Rayner-Canham, Marelene F., and Geoffrey W. Rayner-Canham. *A devotion to their science: pioneer women of radioactivity.* Philadelphia, PA: Chemical Heritage Foundation; Montreal: McGill-Queen's University Press, 1997. 307 p. 0941901165; 0941901157 pbk QC15 539.7 98-131383

Features 17 chapter-length biographies and six brief accounts (where available information was limited) of the first generation of women working in the field of radioactivity. Except for Marie Curie and Lise Meitner, these women are virtually unknown today. The biographies are well researched and thoughtfully written, including descriptions of the actual scientific contributions of these women. The descriptions are not technical and can be easily read by anyone. The style is straightforward and for the most part factual, although in certain instances the authors point out examples of discrimination against women and make editorial comments on the behavior of the women themselves. Well documented, clearly written, and full of material not readily available elsewhere.

1.077

Rayner-Canham, Marelene F., and Geoffrey W. Rayner-Canham. *Women in chemistry: their changing roles from alchemical times to the mid-twentieth century.* Washington, DC: American Chemical Society/Chemical Heritage Foundation, 1998. 284 p. (History of modern chemical sciences, ISSN 1069-2452) 0841235228; 0841233454 pbk QD20 540 98-3890

A significant attempt to redress the omission of women from standard histories of chemistry; overviews the lives, roles, and contributions of women from the beginning of recorded history (where source material is scanty) to the mid-20th century. The first chapter has two paragraphs on Chinese chemistry, but the focus is on European and North American women. The sheer number of women, particularly in the fields of crystallography, radioactivity, and biochemistry, necessitates a very brief treatment of each: the role she played in her field and some account of her personal life to the neglect of her actual scientific contribution. There is a two-fold attempt to compensate for this problem: to provide comprehensive bibliographies for each woman and a picture of the cultural milieu in which she practiced; to discuss the social and historical forces that may have contributed to her choice of field, the discoveries that she made, or the obstacles to be overcome.

1.078

Robert Burns Woodward: architect and artist in the world of molecules, ed. by Otto Theodor Benfey and Peter J.T. Morris. Philadelphia: Chemical Heritage Foundation, 2001. 470 p. (History of modern chemical sciences, ISSN 1069-2452) 0941901254 QD22.W67 547 00-50876

This collection offers some of Robert Woodward's most seminal and stunning landmark publications in synthetic and physical organic chemistry. Although Woodward (Nobel Laureate in chemistry, 1965) published some 210 papers during his lifetime, only 22 publications are reproduced here. Papers contain important new ideas in synthesis and physical organic chemistry, e.g.: Woodward's Rules; the Octet Rules; Woodward-Hoffmann Rules; and papers on total syntheses, including those of quinine, steroids, strychnine, reserpine, chlorophyll, and vitamin B12. Included is Woodward's remarkable and heretofore unpublished Cope Award Lecture; this lecture, partly semiautobiographical, portrays the history of chemical structure from Woodward's unique point of view. Useful appendixes with complete bibliography of his publications, two-page chronology of his life, and name index.

1.079

Russell, Colin Archibald. *Edward Frankland: chemistry, controversy, and conspiracy in Victorian England.* Cambridge; NY: Cambridge University Press, 1996. 0521496365 QD22.F65 540 95-40319

A thoroughly researched, excellent biography of an underappreciated Victorian chemist draws on newly investigated papers of the chemist and his family. Frankland pioneered organometallic chemistry and gave chemistry the concepts of valency and the chemical bond. Well illustrated, fully referenced and indexed.

1.080

Seaborg, Glenn Theodore, with Eric Seaborg. *Adventures in the atomic age: from Watts to Washington.* NY: Farrar, Straus & Giroux, 2001. 312 p. 0374299911 QD22.S436 327.1 00-49522

Offers accounts of the scientific revolution during the first half of the 20th century and of the concentrated intensity of the Manhattan Project; insights into the personalities that participated in these events are brought to final form after his death. Seaborg's insights into university administration in the 1950s are appropriate 50 years later, and his comparisons of the three presidents (Kennedy, Johnson, and Nixon) under whom he served as chairman of the Atomic Energy Commission are enlightening.

1.081

Schütt, Hans-Werner. *Eilhard Mitscherlich, prince of Prussian chemistry*; transl. by William E. Russwy. Washington, DC: American Chemical Society and the Chemical Heritage Foundation, 1997. 239 p. (History of modern chemical sciences) 0841233454 QD22.M58 540 96-33322

Describes the life and work of Mitscherlich (1794-1863) both chronologically and thematically. Mitscherlich came to chemistry from Oriental languages and medicine, and discovered and applied the isomorphism of crystals, impressing Berzelius, who became his lifelong mentor and advocate. Mitscherlich did significant work with benzene, catalysis in ether synthesis, fermentation, and optical rotation of sugars. He was a creative researcher, but cautious in interpretation. He wrote an influential chemistry textbook and was interested in applications of chemistry in industry and medicine.

1.082

Thackray, Arnold, and Minor Myers Jr. *Arnold O. Beckman: one hundred years of excellence.* Foreword by James D. Watson. Philadelphia: Chemical Heritage Foundation, 2000. 379 p. 1 CD-ROM 0941901238 pbk Q185 338.7 00-27014

Beckman was a blacksmith's son who grew up to play a pivotal role in the instrumentation revolution, dramatically changing science, technology, and society. He invented the pH meter and the DU spectrophotometer. The accompanying CD-ROM offers personal views of his life and philosophies.

1.083

Women in chemistry and physics: a biobibliographic sourcebook, ed. by Louise S. Grinstein, Rose K. Rose, and Miriam H. Rafailovich. Foreword by Lilli S. Hornig. Westport, CT: Greenwood Press, 1993. 721 p. 0313273820 QD21 540 92-40224

The incompatibility of women and science is a persistent belief. Compelling stories of women who have worked either in or around a male-dominated field, making substantial contributions in research, education, and scholarly publications. The editors have compiled a representative collection of biobibliographies of women in the physical sciences spanning some three centuries. Entries are arranged alphabetically and follow a three-section format: Biography, telling the individual's life story; Work, detailing the significant contributions made to her discipline; and Bibliography, listing works by and about each woman. Appendixes include biographical, discipline, and chronological information.

1.084

Zewail, Ahmed H. *Voyage through time: walks of life to the Nobel Prize.* Cairo, Egypt; NY: American University in Cairo Press, 2002. 287 p. 9774246772 QD22.Z49 2002-332902

Zewail received the 1999 Nobel Prize in Chemistry "for his studies of the transition states of chemical reactions using femtosecond spectroscopy." Essential information about the award is available at the Nobel Web site. This book is Zewail's autobiography; he grew up in Egypt, earned BS and MS at Egypt's Alexandria University, earned his PhD at the University of Pennsylvania, and is currently professor of chemistry and of physics at the California Institute of Technology, where he holds the Linus Pauling Chair. The book tells the story of his life, his family, his scientific accomplishments, the Nobel award, and his broader cultural reflections, bridging East and West. Thus, the last chapter is titled, "Walks to the Future: My Hope for Egypt and America."

HISTORY OF CHEMISTRY

1.085

Archaeological chemistry: organic, inorganic, and biochemical analysis, ed. by Mary Virginia Orna. Washington, DC: American Chemical Society, 1996. 459 p.

Basic Chemistry

(ACS symposium series, ISSN 0097-6156; no. 625) 0841233950 CC79.C5 930.1 96-4812

Four other ACS symposia on this subject precede this work (*Archaeological Chemistry*, ed. by C.W. Beck, 1974; *Archaeological Chemistry II*, ed. by G.F. Carter, 1978; *Archaeological Chemistry III*, ed. by J.B. Lambert, 1984; and *Archaeological Chemistry IV*, ed. by R.O. Allen, 1989). Papers discuss synthetic and natural glass, metals, soils, textile fibers and dyes, bone and teeth, DNA, and carbon dating. Analysis methods range from wet chemical methods to advanced instrumental analysis.

1.086

Berson, Jerome A. *Chemical creativity: ideas from the work of Woodward, Hückel, Meerwein, and others.* Weinheim; NY: Wiley-VCH, 1999. 195 p. 3527297545 QD11 540 99-206238

Contents: Introduction; Discoveries missed, discoveries made: two case studies of creativity in chemistry; Erich Hückel and the theory of aromaticity; Reflections on theory and experiment; The dienone-phenol mysteries; Meditations on the special convictive power of symmetrization experiments; Epilogue.

1.087

Brock, William H. *The Norton history of chemistry.* NY: W.W. Norton, 1993 (c1992). 744 p. 0393035360; 0393310434 pbk QD11 540 93-19054

Concisely covers early chemistry, up to the end of the 18th century, in the first two chapters. Subsequent chapters discuss the Lavoisier revolution and Dalton's work; the progress of chemistry in the 19th century including the complex process of unravelling organic structures; the growth of scientific industrial chemistry; the emergence of the Periodic Law; the beginnings of physical chemistry; chemical education and early periodicals; the history of modern chemical bond theory; organic structure and mechanism; rebirth of inorganic chemistry; and modern chemical industry. Valuable bibliographic essay; brief appendix on museums and collections; full index.

1.088

A Century of chemistry: the role of chemists and the American Chemical Society. Herman Skolnik, chairman, board of editors; Kenneth M. Reese, ed. Washington: The Society, 1976. 468 pp. 0841203075 QD1 540.6 76-6126

Pt. I: historical perspectives, chemical education, professionalism, publications, impact of government, public affairs, intersociety relations, governance, headquarters staff and operations, and ACS divisions and their disciplines. Pt. II: the record: a list of all ACS presidents, personnel down through local section chairmen, membership statistics, portraits, etc.

1.089

Chemical sciences in the modern world, ed. by Seymour H. Mauskopf. Philadelpha: University of Pennsylvania Press, 1993. (The Chemical sciences in society series) 0812231562 QD11 540 93-30055

About 20 papers by well-known chemists highlighting historical advances in chemistry, from the 18th century to the present.

1.090

Chymia. Editor in chief, Henry M. Leicester. American Chemical Society, Division of the History of Chemistry; Edgar F. Smith Memorial Collection, University of Pennsylvania. v. 1-12, 1948-1967. ISSN 0095-9367 QD11 540.9 48-7051

Annual studies in the history of chemistry. Essays (12 or so) in each volume.

1.091

Classics in the theory of chemical combination, ed. by O. Theodor Benfey. Malabar, FL: R.E. Krieger Pub. Co., 1981, c1963. 191 p. 0898743680 QD455 541 81-8300

Originally published by Dover in 1963 (Classics of science, v. 1). Chapters on valence theory and development by internationally famous chemists Wöhler, Liebig, Laurent, Williamson, Frankland, Kekulé, Couper, van't Hoff, and Le Bel.

1.092

Cobb, Cathy, and Harold Goldwhite. *Creations of fire: chemistry's lively history from alchemy to the atomic age.* NY: Plenum Press, 1995. 475 p. 0306450879 QD11 540 95-24804

Discusses the beginnings of chemistry up to 1700 and the phlogiston movement, then moves from Lavoisier and the French Revolution to the discovery and synthesis of the modern-day superheavy elements. Detailed chapter notes and bibliography.

1.093

The Development of chemistry, 1789-1914, ed. by David Knight. London; NY: Routledge/Thoemmes Press, 1998. 10 v. 0415179122 (set) QD18.E85 540 98-4837

This set reproduces original articles and books written between 1789-1914 describing the techniques and findings of chemistry's inception as a recognized science. Volume 1 collects 38 short, readable papers, many transcribed from lectures, the earliest published in *Philosophical Transactions of the Royal Society* and the Scottish *Annals of Philosophy*, and later, in *Proceedings of the Royal Institution*. The authors are already well known to chemists, e.g., H. Davy, D. Mendeleev, J. Dewar, Lord Rayleigh, J.J. Thomson, S. Arrhenius, and W.H. Bragg. The forces behind chemical reactions and the experiments used to determine them are described, especially at the interface of chemistry and physics. The character and personality of the authors come through from their writings and provide an added dimension to reading these original papers. V. 1 also includes articles on a variety of topics, from problems of water pollution (Frankland), to the properties of diamonds (Crookes), to the very practical synthetic organic chemistry used in the manufacture of dyes (mauve, magenta, and indigo) (Hofmann). V. 2, Elements of Chemistry, by Antoine Lavoisier, and v. 3, Researches into the Laws of Chemical Affinity, by C.L. Berthollet, were used to transmit the chemical knowledge of the time. Because they are reproduced exactly from the first English translations, they are sometimes difficult to read but are accessible to anyone, and descriptions and diagrams of experiments are wonderful reading and demonstrate the inventiveness of these early chemists. V. 4 and 5, by Berzelius and Faraday, offer practical advice and describe how to do the experiments and how to design instruments for a particular purpose. V. 6, a wonderful text by J. Liebig, describes organic analysis methods with diverse applications (i.e., physiology, agricultural chemistry). The language of chemical formulae (as we know them) is described in v. 7 by A. Laurent, and v. 8 is by A. Wurtz, a German chemist who used atoms in formulating chemical theories, as opposed to the influential French chemist of the time, M. Berthelot, who preached their hypothetical existence. V. 9 treats spectral analysis by N. Lockyer, an amateur scientist who first studied solar spectral lines. Last is a small work by J.H. van't Hoff, "The Arrangement of Atoms in Space," in which many of the concepts that form the basis of chemical structure are discussed (e.g., stereochemistry, the tetrahedral carbon atom).

1.094

Donovan, Arthur. *Antoine Lavoisier: science, administration, and revolution*. Oxford; Cambridge, MA: Blackwell, 1993. 351 p. (Blackwell science biographies) 0631178872 QD22.L4 540 93-19063

Includes biographical and historical episodes; scientific texts; and institutions. Full references to the most recent literature on Lavoisier's scientific investigations and on his major administrative activities in the French government. Eight well-chosen illustrations; good index.

1.095

Duncan, Alistair M. *Laws and order in eighteenth-century chemistry*. Oxford: Clarendon Press; NY: Oxford University Press, 1996. 253 p. 0198558066 QD18.E85 540 96-165706

This book centers on the development of chemical affinity during the 18th century, set in the context of the preceding century. Some 19th-century connections are also made, such as Dalton's atomic theory and Guldberg and Waage's law of mass action. Discussion of an extensive set of tables of chemical affinity begins with Geoffroy, and includes phlogiston theory, part of the 18th-century environment, finding its way into affinity tables. Strong emphasis on chemical affinity and chemistry's move toward intellectual autonomy.

1.096

Enlightenment science in the romantic era: The chemistry of Berzelius and its cultural setting, ed. by Evan M. Melhado and Tore Frängsmyr. Cambridge; NY: Cambridge University Press, 1992. 246 p. (Uppsala studies in history of science, v. 10) 0521417759 QD22 540 91-45939

Nine essays by nine authors on various aspects of Berzelius's life and work, especially on his work to lay the stoichiometric foundation for chemistry.

1.097

Ferry, Georgina. *Dorothy Hodgkin: a life*. Cold Spring Harbor, NY: Cold Spring Harbor Laboratory, 2000 (c1998). 423 p. 0879695900 QD903.6H63 548 00-22666

Dorothy Crowfoot Hodgkin was the only female British scientist to win a Nobel Prize. Traces the life of this accomplished but not very well known woman, particularly with respect to the problems she encountered as a woman scientist in a field dominated by men; explains her scientific achievements; and describes her active involvement with the political and social issues of the 20th century. Sufficient account of her role as an X-ray crystallographer in increasing the understanding of insulin, penicillin, and vitamin B.

1.098

Garfield, Simon. *Mauve: how one man invented a color that changed the world.* NY: W.W. Norton, 2001. 222 p. 0393020053 TP140.P46 666 00-69533

William Henry Perkin and his color stained the 19th century like nothing else. Dyes were Victorian "high tech," opening the age of synthetic chemistry and industry. There is a solid opening chapter on Perkin's famous visit to the US the year before his death but nothing of the splash dyes made at London's Great Exhibition (1851) and the famous St. Louis World's Fair (1904). A chapter on toxicity and environmental impacts suggests concerns a century before our present preoccupation.

1.099

Greenberg, Arthur. *A chemical history tour: picturing chemistry from alchemy to modern molecular science.* NY: John Wiley & Sons, 2000. 312 p. (A Wiley-Interscience publication) 0471354082 QD11 540 99-38865

Nine sections—Practical Chemistry, Mining and Metallurgy; Spiritual and Allegorical Alchemy; Iatrochemistry and Spagyricall Preparations; Chemistry Begins to Emerge as a Science; Modern Chemistry Is Born; Chemistry Begins to Specialize and Helps Farming and Industry; Teaching Chemistry to the Masses; The Approach to Modern Views of Chemical Bonding; and Postscript— offer a fascinating tour through the history of chemistry.

1.100

Hudson, John. *The history of chemistry.* NY: Chapman & Hall, 1992. 285 p. 041203641x; 0412036517 pbk QD11 540 92-8311

Contents: Early processes and theories; Alchemy; From alchemy to chemistry; Phlogistic and pneumatic chemistry; Lavoisier and the birth of modern chemistry; The chemical atom; Electrochemistry and the dualistic theory; The foundation of organic chemistry; The Karlsruhe Congress and its aftermath; Organic chemistry since 1860; Atomic structure, radiochemistry and chemical bonding; Inorganic chemistry; Physical chemistry; Analytical chemistry; Chemistry and society.

1.101

Ihde, Aaron John. *The development of modern chemistry.* NY: Harper & Row, 1964; Reprinted: NY: Dover, 1984. 851 p. 0486642356 pbk QD11 540 64-15152; 82-18245 (1984 ed.)

In 27 chapters, 4 appendixes, notes, name and subject indexes. Pt. I: Foundations of chemistry; Pt. II: Period of fundamental theories; Pt. III: Growth of specialization; Pt. IV: Century of the electron.

1.102

Instruments and experimentation in the history of chemistry, ed. by Frederic L. Holmes and Trevor H. Levere. Cambridge, MA: MIT Press, 2000. 415 p. (Dibner Institute studies in the history of science and technology) 0262082829 QD53 542 99-21064

An important addition to the literature of chemistry. The 14 chapters by recognized scholars, two serving as editors, are divided into three chronological groups: three on alchemy, six on the 18th century, and five on the 19th and 20th centuries. Introductions to the sections offer excellent perspective and coherence among the varied emphases. Chapter notes; some have bibliographies. Quotations and illustrations of apparatus are included.

1.103

Knight, David M. *Ideas in chemistry: A history of the science.* New Brunswick, NJ: Rutgers University Press, 1992. 213 p. 0813518350; 0813518369 pbk QD11 540 91-47598

A collection of essays with 460 notes and references.

1.104

Krebs, Robert E. *The history and use of our earth's chemical elements: a reference guide.* Westport, CT: Greenwood Press, 1998. 346 p. 0313301239 QD466 546 96-49735

For most elements, the information provided is clear-cut and thorough, although it was printed too soon to have the names for elements 104-109 (assigned in fall of 1997 by the International Union of Pure and Applied Chemistry). Limited information addressing biological roles of elements (readers will need to see the bioinorganic chemistry literature).

1.105

Lambert, Joseph B. *Traces of the past: unraveling the secrets of archaeology through chemistry.* Reading, MA:

Addison-Wesley, 1997. 319 p. (Helix books) 0201409283
CC75 930.1 97-11454

The best book on archaeological chemistry in print. Written in a storyteller's fluid prose, it is a rigorous, reliable, and learned account of the contributions of chemistry to archaeology. Chapters are arranged by materials: stone, soils, pottery, colorants, glass, metals, organic materials, and human remains. Examples are drawn from all cultures and all time periods. Copiously illustrated with graphs, maps, simple chemical equations, black-and-white photographs, and 16 color plates; glossary of nearly 300 technical terms; extensive bibliography, arranged by chapter and section. Thorough analytical index.

1.106

Levere, Trevor H. *Transforming matter: a history of chemistry from alchemy to the buckyball.* Baltimore: Johns Hopkins University Press, 2001. 215 p. (Johns Hopkins introductory studies in the history of science) 080186609x; 0801866103 pbk QD11 540 00-11487

This book treats the transformations that have occurred in the study of chemistry, i.e., the secretiveness of alchemy was replaced by quantitative experiment in the 18th century, and theoretical studies moved to a position of greater importance in the 20th century. This complex story revolves around some important themes: classification of elements and compounds, relationships between theory and applications, and the changing nature of chemistry. Twentieth-century chemistry is covered quite briefly.

1.107

Mason, Stephen Finney. *Chemical evolution: origins of the elements, molecules, and living systems.* Oxford: Clarendon Press; NY: Oxford Univ. Press, 1991. 317 p. 0198552726 QH325 577 90-7321

Covers the history of chemistry, astrophysics, geology, and biochemistry: 14 chapters range from atomic structure to the origin of the universe/earth to bioenergetics-bioinformation. Discusses radioactivity, distribution of elements; stellar nucleosynthesis; catalysis, glycolysis and respiration, chemiosmosis, and photosynthesis; proteins, nucleic acids and genetic code relationships; and other related material.

1.108

Nye, Mary Jo. *From chemical philosophy to theoretical chemistry: dynamics of matter and dynamics of*

disciplines, 1800-1950. Berkeley: University of California Press, 1993. 328 p. 0520082109 QD452 541 92-43114

Three main themes are discussed: The demarcation of chemistry and physics as separate disciplines; The development of physical chemistry as a subdiscipline of chemistry and its role in the relationship of chemistry to physics; and The application of physical chemical principles, especially those of thermodynamics and kinetics, to organic chemistry in the development of the theory of reaction mechanisms.

1.109

Partington, James Riddick. *A history of chemistry.* London: Macmillan; NY: St. Martin's, 1961-70. v.1-4 in 4 volumes. QD11 540 62-1666

V. 1, pt. 1, 1970 titled Theoretical background. Very scholarly; the four volumes are arranged as follows: I. Earliest period to 1500 AD. II. 1500-1700. III. 1700-1800. IV. 1800-present. Heavily footnoted and referenced.

1.110

Partington, James Riddick. *A short history of chemistry.* NY: The Macmillan Company, 1937. 386 p. QD11 540.9 38-27060

Bibliography: p. 364-368. 3rd ed., rev. and enlarged, 1957. Reprinted, NY: Dover, 1989. Abbreviated version of the four-volume set, previously cited.

1.111

Salzberg, Hugh W. *From caveman to chemist: circumstances and achievements.* Washington, DC: American Chemical Society, 1991. 294 p. 0841217866; 0841217874 pbk QD11 540 90-33612

Chapters treat ancient technology; Hellenic chemical science; the Hellenistic and Roman eras; Islamic alchemy; Medieval and Renaissance European artisans, alchemists and natural philosophers, medicine and medical alchemy; The information explosion; Changing the frame of reference: the undermining of the scientific establishment; The first chemists; Phlogiston; Lavoisier and the chemical revolution; Atomic weights and molecular formulas; Atomic arrangements; The divisible atom.

1.112

Stillman, John Maxson. *The story of alchemy and*

early chemistry. NY: Dover Publications, 1960. QD11 540.1 60-3183

An unabridged and unaltered republication of the first edition, 1924, which appeared under the title, *The Story of Early Chemistry*. Covers early chemistry up through Lavoisier.

1.113

Weeks, Mary Elvira. *Discovery of the elements*. 7th ed., completely rev. and new material added by Henry M. Leicester. Easton, PA: Journal of Chemical Education, 1968. QD466 546 68-15217

(Information here is from 5th ed, rev. and enl., 1945.) Many illustrations and portraits. Treats elements known to ancients and alchemists; discovery of phosphorus; 18th-century metals; three important gases; Rutherford; chromium, molybdenum, tungsten, and uranium; discovery of technetium (columbium), tantalum, vanadium; platinum metals; potassium, sodium, lithium; alkaline earths; zirconium, titanium, cerium, thorium; beryllium, boron, silicon, aluminum; spectroscopic discoveries; periodic system of elements; rare earths; halogens; inert gases; naturally radioactive elements; recently discovered elements.

1.114

Williams, R.J.P., and J.J.R. Fraústo da Silva. *Bringing chemistry to life: from matter to man*. Oxford; NY: Oxford University Press, 1999. 548 p. 0198505469 QD33 540 99-24780

The third volume of a "trilogy" by these authors; the first two volumes were titled: *The Biological Chemistry of the Elements* (1991) and *The Natural Selection of the Chemical Elements* (1996). The book presents the principles of chemistry, in a hierarchical structure designed for understanding cellular processes in terms of the most basic properties of matter. Of great value to teachers.

Chemistry Today

1.115

Atkins, P.W. *The periodic kingdom: a journey into the land of the chemical elements*. NY: Basic Books, 1995. 161 p. (Science masters series) 0465072658 QD466 541.2 95-7362

This delightful book uncovers the richness and centrality of the periodic table for the nontechnical reader, but even the expert will find new and enjoyable insights into the table, or in the author's allegory, the "periodic kingdom." Although addressed to the scientifically uninitiated, this is a book for the well educated.

1.116

Ball, Philip. *Stories of the invisible: a guided tour of molecules*. Oxford; NY: Oxford University Press, 2001. 204 p 0192802143 QD461 541.2 2001-36601

Ball (with degrees in chemistry and physics) suggests exploring the world of "molecules," a term as yet untarnished; he argues, let these molecules be the molecules of life, to demonstrate that chemistry is not a science we do, but rather the science we are. What results is a fascinating journey through biochemistry that serves as a springboard for brief forays into materials science and nanotechnology. The systems considered are necessarily complicated. Ball finds analogies to make them accessible to the general reader, while experts can still admire their beauty.

1.117

Breslow, Ronald. *Chemistry today and tomorrow: the central, useful, and creative science*. Washington, DC: American Chemical Society; Boston: Jones and Bartlett, 1997. 134 p. 0841234604 (ACS); 0763704636 (Jones and Bartlett) QD31.2 540 96-45078

What do chemists do? What impact does chemistry have on our lives and our society? The selection of topics is timely, with particular attention given to medicinal and environmental chemistry. The author makes the important point that all materials are chemical, and that natural products include some of the most potent toxins known.

1.118

Emsley, John. *The elements*. Oxford: Clarendon Press; NY: Oxford Univ. Press, 1998. 3rd ed. 292 p. 0198558198; 019855818x pbk QD466 546 98-213249

Two densely packed pages of information about each element, including physical, chemical, and biological data. New information on health hazards, mineral sources, electron binding energies, and elements discovered. Brief chapters on the history of the periodic table, and the discovery and the abundance of the elements. Brief index.

1.119

Emsley, John. *Molecules at an exhibition: portraits of* *intriguing materials in everyday life.* Oxford; NY: Oxford Univ. Press, 1998. 250 p. 0198502664; 0198503792 pbk QP514.2 540 97-36754

Describes a wide range of chemical materials intriguing for both scientists and the general public because of their unique applications in everyday life. This volume offers portraits of chemicals grouped into eight galleries, each with a common theme. Examples of these themes include molecules in foods, metals, molecules that help and harm the young, detergents, fuels, poisons, materials, and environmental contaminants. The portraits then describe the molecules in language that the general reader can understand and follow, with application and analysis of potential riches and benefits.

1.120

Emsley, John. *Nature's building blocks: an A-Z guide to the elements.* Oxford; NY: Oxford University Press, 2001. 538 p. 0198503415 QD466 546 2002-283442

Each chapter is dedicated to a single element, arranged alphabetically; some are short while others are very extensive, a reflection of the element's abundance and importance. Pronunciations are provided, as are name variations in several languages and the origins of the element names. Discussed are history of the element's discovery, use in the human body, military applications, economic and environmental applications, key chemical properties, and other features.

1.121

Hoenig, Steven L. *Basic training in chemistry.* NY: Kluwer Academic/Plenum Publishers, 2001. 184 p. 0306465469 QD31.2 540 00-67108

Gathers together essential information on general, inorganic, organic, and instrumental analysis as a supplemental source of material for students as well as practitioners.

1.122

Hoffmann, Roald. *The same and not the same.* NY: Columbia University Press, 1995. 294 p. (The George B. Pegram lecture series) 02311001384; 0231101392 pbk QD37 540 94-39457

The book originated as a series of lectures at Brookhaven National Laboratory in 1990. The central theme is the importance of molecular shapes; topics include the thalidomide episode, the social responsibility of scientists, a philosophical discussion of reductionism, environmental

issues, and the life of Fritz Haber. Excellent photographs; Hoffman's writing style is enjoyable.

1.123

Karukstis, Kerry K., and Gerald R. Van Hecke. *Chemistry connections: the chemical basis of everyday phenomena.* 2nd ed. San Diego: Academic Press, 2003. 250 p. 0124001513 pbk QD37 540 2003-102993

A compilation of a very significant collection of practical chemical facts, with concise descriptions. The book ranges through such diverse subjects as medicine, home settings, outer space, office settings, and the arts. Two major cases of possible danger are discussed—ammonia and bleach mixtures in the home and the health concern associated with the production of artificial fog and smoke in the performing arts. Keywords; references; relevant Web site addresses (1st ed. 2000).

1.124

Lewis, Grace Ross. *1001 chemicals in everyday products.* 2nd ed. NY: Wiley, 1999. 388 p. (A Wiley-Interscience publication) 0471292125 pbk TP200 363.17 98-6419

A second edition (1st ed., 1994) of a monograph that was written to "answer questions and provide information about all the chemicals that you come in contact with every day." Part 1 provides brief, clearly written answers to some 200 questions about food additives, cosmetics, cleaning products, and other materials. Part 2 contains an alphabetical listing of hundreds of substances; for each there is a brief but clearly written summary of uses, precautions associated with uses, and a valuable list of synonyms. Appendix listing locations of Poison Control Centers in the US and Canada; extremely detailed index.

1.125

Newton, David E. *Recent advances and issues in chemistry.* Phoenix, AZ: Oryx Press, 1999. 294 p. (Oryx frontiers of science series) 1573561606 QD31.2 540 98-40880

Aimed at readers with an interest in chemistry and focused on advances in chemical research during 1996-98. Newton summarizes research findings in a variety of areas and then turns to social issues involving chemistry and chemical technology. Global warming and ozone layer depletion are discussed, together with the less-familiar problems of food safety and control of chemical weapons. The second half contains biographical sketches of chemists who have made important research contributions, the ozone layer, the Delaney clause, and chemical weapons. Chapters on the future of chemistry, on careers in chemistry, and on recent trends in the chemical industry. Listings of scientific organizations; information on print and electronic resources; glossary of chemical terms.

1.126

Of minds and molecules: new philosophical perspectives on chemistry, ed. by Nalini Bhushan and Stuart Rosenfeld. Oxford; NY: Oxford University Press, 2000. 299 p. 0195128346 QD6 540 99-40329

An anthology devoted to work in the philosophy of chemistry. Essays were prepared by both chemists and philosophers, adopt distinctive philosophical perspectives on chemistry, and collectively offer both a conceptualization of and justification for this field.

1.127

Partnerships in chemical research and education, ed. by James E. McEvoy. Washington, DC: American Chemical Society, 1992. 159 p. (ACS symposium series; 478) 0841221731 QD40 540 91-32420

A set of 15 papers delivered at two ACS symposia in 1990 describing programs developed during the past 20 years establishing partnerships between universities and industries.

1.128

Stimulating concepts in chemistry, ed. by Fritz Vögtle, J. Fraser Stoddart, and Masakatsu Shibasaki. Weinheim; NY: Wiley-VCH, 2000. 396 p. 3527299785 QD39 540 2001-274155

A collection of 24 essays by leading researchers, overviewing the most recent advances in their fields. Included are essays on self-assembly, nanochemistry, molecular machines, super-size biomolecules, and new materials.

1.129

Taber, Keith. *Chemical misconceptions: prevention, diagnosis and cure*. Cambridge: Royal Society of Chemistry, 2002. 2 v. 085404390x (set) QD40 540.71241

Volume 1 treats the theories of chemistry that may cause problems for students; v. 2 provides strategies for coping with misconceptions that students have.

1.130

What every chemist should know about patents, prepared by the Committee on Patents and Related Matters. Washington, DC: American Chemical Society, available as a .pdf file, online at http://www.chemistry.org/portal/resources/ACS/ACSContent/government/publications/Chem_Patent2001.pdf. 21 p.

This booklet is intended for chemists in industry, small businesses, and academe, prompted by enactment of legislation in 1995 that introduced major changes in U.S. patent laws. Included is an overview of the obtainment, use, and value of patents in an easily understood form. Discusses patent essentials and patenting procedures.

THE FUTURE OF CHEMISTRY

1.131

Chemistry at the beginning of the third millennium: molecular design, supramolecules, nanotechnology, and beyond; proceedings of the German-Italian meeting of Coimbra Group universities, Pavia, 7-10 October 1999. Luigi Fabbrizzi and Antonio Poggi, editors. Berlin; NY: Springer, 2000. 342 p. 3540674608 QD1 540 00-41048

These 15 chapters describe some of the most intriguing and promising topics in modern chemistry. Authors describe how the research will develop in the next decade.

1.132

The new chemistry; ed. in chief, Nina Hall. Cambridge, UK; NY: Cambridge University Press, 2000. 493 p. 0521452244 QD39 540 99-16729

Showcases chemical research done over the last 30 years, with leading chemists writing 17 chapters on their fields of expertise to present their work to readers ranging from chemists to students, teachers, scientists in other disciplines, to anyone interested in the progress of science.

Chapter authors include Roald Hoffmann, Ilya Prigogine, Glenn Seaborg, and Gabor Somorjai.

1.133

Royal Society of Chemistry (Great Britain). *Cutting edge chemistry*, compiled by Ted Lister. London: Royal Society of Chemistry, 2000. 196 p. 0854049142 QD42

Explains some of the extraordinary developments in chemistry during the 20th century, and describes some that are likely to happen in the future, including technological applications, practical aspects, and biological breakthroughs.

SAFETY LITERATURE

1.134

Armour, M.A. *Hazardous laboratory chemicals disposal guide.* 3rd ed. Boca Raton, FL: Lewis Publishers, 2003. 557 p. 1566705673 QD64 542 2002-43358

Contains entries for more than 300 chemicals, including 15 pesticides used in greenhouses. Emphasis is on disposal methods that turn hazardous wastes into nontoxic materials, including acid-base neutralization, oxidation/reduction, and precipitation of toxic ions as insoluble salts. Hazardous reactions are mentioned, as are physiological responses to the waste materials.

1.135

Concise manual of chemical and environmental safety in schools and colleges. The Forum for Scientific Excellence, Inc. Philadelphia: Lippincott, 1988-1991. 5 vol. (Chemical and environmental safety and health in schools and colleges series) 0397530072 (v. 1) T55.3 371.7 89-135275

V. 1: Basic principles; v. 2: Hazardous chemicals classes; v. 3: Chemical interactions; v. 4: Safe chemical storage; v. 5: Safe chemical disposal.

1.136

CRC handbook of laboratory safety, ed. by A. Keith Furr. 5th ed. Boca Raton, FL: CRC Press, 2000. 774 p. 0849325234 QD51 542 99-86773

Covers responsibility assignment, emergency programs, laboratory facilities (design and equipment), laboratory operations, nonchemical laboratories, and personal protective equipment. New topics in this edition include AIDS and laboratory infections, recombinant DNA, carcinogenic and teratogenic effects, regulatory requirements, medical surveillance programs, and laws and regulations relating to animal care and use.

1.137

Hajian, Harry G. Sr., and Robert L. Pecsok. *Working safely in the chemistry laboratory.* Washington, DC:

American Chemical Society, 1994. 277 p. 0841227063; 0841227071 pbk QD63.5 542 93-45353

Includes useful information for undergraduate chemistry laboratory students or chemistry technicians. Covers a broad spectrum of laboratory-related topics without much detail on any one subject, and is not a definitive reference; there are short lists of references at the end of some chapters. Discusses first aid, safe practices in the laboratory, toxicity of chemicals, fire safety, electrical and radiation hazards, treatment of data, and the use of laboratory tools.

1.138

Hall, Stephen K. *Chemical safety in the laboratory.* Boca Raton, FL: Lewis Publishers, 1994. 242 p. 0873718968 TP187 660 93-28653

This guidebook for implementing and managing a chemical hygiene program begins with the Occupational Safety and Health Act of 1970, explains federal and state safety regulations, and outlines the responsibilities of workers, managers, safety instructors, and laboratory users. Discusses regulations pertaining to chemical handling procedures, inventory control, labeling systems, safety equipment, monitoring of chemical exposure, emergency response planning, and record keeping.

1.139

Hazards in the chemical laboratory, ed. by S.G. Luxon. 5th ed. Cambridge: Royal Society of Chemistry, 1992. 675 p. 0851862292 QD63.5 542 93-121147

The previous (4th) edition was edited by Leslie Bretherick. Discusses various aspects of chemical safety and offers pertinent chemical information about specific laboratory compounds. Includes legislation, safety planning, laboratory design, fire protection, reactive chemical hazards, toxicology, health hazards, radiation safety, electrical hazards, and US chemical laboratory safety standards.

1.140

Improving safety in the chemical laboratory: a practical guide, ed. by Jay A. Young. 2nd ed. NY: Wiley, 1991. 406 p. (A Wiley-Interscience publication) 0471530360 QD51 542 90-24189

Offers advice on keeping labs safe: recognizing "close calls", eliminating accidents, using personal protective equipment, establishing training programs, and utilizing effective emergency procedures.

Basic Chemistry

1.141

Laboratory waste management: a guidebook, written by ACS Task Force on Laboratory Waste Management. Washington, DC: ACS, 1994. 211 p. 0841227357; 0841228493 pbk TD899.L32 660 93-45546

An easily accessible resource for developing effective strategies for managing laboratory wastes—from the laboratory perspective. Chapters include laws and regulations, organizational responsibilities, worker training, identification and characterization of wastes, waste reduction and handling, monitoring and control, and regulations. Useful glossary and appendixes.

1.142

Lefèvre, Marc J. *First aid manual for chemical accidents.* 2nd English-language ed. rev. by Shirley A. Conibear. NY: Van Nostrand Reinhold, 1989. 261 p. 0442204906 RC963.3 615.9 88-27996

Valuable information about toxic chemical exposure, this book includes a 16-page index of nearly 500 chemicals. Toxicology information and overexposure symptoms are included.

1.143

Less is better: laboratory chemical management for waste reduction, prepared by the Task Force on Laboratory Waste Management, 2nd ed. Washington, DC: American Chemical Society, Dept. of Govt. Relations and Science Policy, 1993. 24 p. QD51

1.144

Prudent practices in the laboratory: handling and disposal of chemicals. Committee on Prudent Practices for Handling, Storage, and Disposal of Chemicals in Laboratories, Board on Chemical Sciences and Technology, Commission on Physical Sciences, Mathematics, and Applications, National Research Council. Washington, DC: National Academy Press, 1995. 427 p. 0309052297 T55 660 95-32461

Combines the older volumes *Prudent Practices for Handling Hazardous Chemicals in Laboratories* and *Prudent Practices for Disposal of Chemicals from Laboratories.* Addresses potential hazards including fire, explosion, and toxics. Offers recommendations for developing a laboratory safety program. Interprets the complex legislation and current regulations on hazardous waste handling. Proposes procedures for waste chemical disposal and for developing an acceptable waste management system for the lab.

1.145

Safety in academic chemistry laboratories. 6th ed. Washington, DC: American Chemical Society, 1995. 80 p. 0841232598 QD63.5 542 97-101932

A publication of the ACS Committee on Chemical Safety. Guides both students and faculty in establishing safe laboratory practices. Helpful reference section.

1.146

Saunders, G. Thomas. *Laboratory fume hoods: a user's manual.* NY: Wiley, 1993. 123 p. (A Wiley-Interscience publication) 0471469356 QD54 542 92-39024

Treats the fume hood, fume hood systems, qualitative and quantitative testing, OSHA regulations, and hood specifications.

1.147

Stricoff, R. Scott, and Douglas B. Walters. *Handbook of laboratory health and safety.* 2nd ed. NY: Wiley, 1995. 462 p. (A Wiley-Interscience publication) 047102628x QD51 001.4 94-19358

Reviews safe workplace methods and compliance with OSHA, EPA, and FDA regulations (rev. ed. of *Laboratory Health and Safety Handbook,* by R. Scott Stricoff and Douglas B. Walters, 1990).

1.148

Waste disposal in academic institutions, ed. by James A. Kaufman. Chelsea, MI: Lewis Publishers, 1990. 192 p. 0873712560 QD64 363.72 90-5400

Papers presented at a symposium held at the Third Chemical Congress of North America, Toronto, Canada, 1988. Includes federal regulations for waste disposal as related to academic institutions; academic waste disposal programs; identification of unknown chemicals; methods for treating and handling wastes; and waste disposal practices.

1.149

Young, Jay A., Warren K. Kingsley, and George H. Wahl Jr. *Developing a chemical hygiene plan.* Washington, DC: American Chemical Society, 1990. 58 p. 0841218765 QD51 542 90-46721

This concise presentation will help colleges and universities as well as other organizations to develop a chemical hygiene plan that is in accordance with federal regulations. Useful appendixes, including OSHA's Laboratory Standard.

Problem Manuals, Educational Materials, and Computer Materials

1.150

Alley, Michael. *The craft of scientific presentations: critical steps to succeed and critical errors to avoid.* NY: Springer, 2003. 241 p. 0387955550 pbk Q223 808 2002-30237

Provides examples from contemporary and historical scientific presentations to show what makes an oral presentation effective. Four perspectives are offered: speech, structure, visual aids, and delivery.

1.151

Billo, E. Joseph. *Excel for chemists: a comprehensive guide.* 2nd ed. NY: Wiley-VCH, 2001. 483 p. 1 computer disk. 0471394629 pbk QD39.3.S67 542 2001-24022

An excellent reference for applying Excel to chemical calculations, from the most simple to the highly complex. Describes the basic operation of Excel and its functions related to recording, summarizing, and graphing so someone with a basic background in science and chemistry can follow. Knowledge of research methods is required to understand and input experimental data directly into Excel. Examples of useful applications of Excel's computing and graphical capabilities are found throughout, as well as examples of the treatment of solution equilibria, spectrophotometric, binding constant, and kinetic data.

1.152

Diamond, Dermot, and Venita C.A. Hanratty. *Spreadsheet applications in chemistry using Microsoft Excel.* NY: Wiley, 1997. 244 p. 1 computer disk (A Wiley-Interscience publication) 0471140872 pbk QD39.3.S7 542 96-26223

This book instructs readers in use of Microsoft Excel spreadsheet program (chapters 1 and 3), and spreadsheet applications in chemistry. The accompanying disk contains spreadsheet files discussed in the text. Chapters go through the basics of Excel, focusing mainly on the scientific uses of a spreadsheet, and present graphing, statistical methods, regression analysis, and the use of macros. Case studies treat quantum chemistry, kinetics, equilibria, titration methods, and importing experimental data from instrumental output. Detailed Excel instructions and some explanation of the underlying chemistry are provided for all case studies.

1.153

Goldberg, David E. *Schaum's outline of theory and problems of beginning chemistry.* 2nd ed. NY: McGraw-Hill, 1999. 366 p. (Schaum's outline series) 0071346694 QD41 540.53 99-23440

Problems (673) with stepwise solutions; 238 additional practice problems with answers; explanations of basic chemical concepts.

1.154

Jurs, Peter C. *Computer software applications in chemistry.* 2nd ed. NY: Wiley, 1996. 291 p. (A Wiley-Interscience publication) 0471105872 QD39.3E46 542 95-41349

Discusses utility software and numerical methods for tasks such as curve fitting, numerical integration, numerical solution of differential equations, and use of pseudorandom numbers, as well as areas of active research in chemistry, such as information handling (for chemical structures, infrared spectra, and mass spectra), molecular mechanics, pattern recognition, and artificial intelligence. The book concludes with graphic methods for displaying data and molecular structures. Includes FORTRAN programs of manageable size. Pertinent chapter references.

1.155

Kenkel, John, Paul B. Kelter, and David S. Hage. *Chemistry: an industry-based introduction with CD-ROM.* Boca Raton, FL: Lewis Publishers, 2001. 520 p. CD-ROM 1566703034 QD33 540 00-30343

Designed to tie traditional chemistry topics to the real world of chemistry in the workplace. Chapters discuss basic chemical concepts, problems with answers, and sidebars that offer a look at industrial situations. The CD-ROM features reviews, demonstrations, interactive quizzes, and applications scenarios.

1.156

Rosenberg, Jerome L., and Lawrence M. Epstein. *Schaum's outline of theory and problems of college chemistry.* 8th ed. NY: McGraw-Hill, 1997. 386 p. (Schaum's outline series) 0070537097 pbk QD41 540 96-46610

Summarizes chemical principles found in a first course through the use of solved problems with solutions. Outdated techniques have been replaced by newer developments, such as those in environmental chemistry, biochemistry, and medicinal chemistry.

1.157

Steiner, Erich. *The chemistry maths book.* Oxford; NY: Oxford University Press, 1996. 542 p. 0198559143; 0198559135 pbk QA37.2 510 95-51779

An excellent quick reference to math concepts ranging from number theory to matrix algebra, with emphasis on applications to chemistry. Exercises at the ends of chapters; answers are included. Historical footnotes add depth. Clearly written with excellent diagrams, and treats subjects, such as the section 9.17 treatment of conversion from Cartesian to polar coordinates, in more detail than usually found in chemistry textbooks.

1.158

Tebbutt, Peter. *Basic mathematics for chemists.* Chichester; NY: Wiley, 1998. 2nd ed. 275 p. 0471972835; 0471972843 pbk QA37.2

This introductory work shows how, where, and why certain mathematical operations are used in practice. Examples, problems, and applications guide the reader from basic mathematical principles through complex numbers, vectors, determinants, and matrices.

1.159

Using computers in chemistry and chemical education, ed. by Theresa Julia Zielinski and Mary L. Swift. Washington, DC: American Chemical Society, 1997. 385 p. 0841234655 QD39.3.E46 542 97-14325

Experts present and discuss many diverse topics to make the reader aware of how computers can be used in a chemistry program. Implementation of any topic mentioned requires further reading; most of the chapters conclude with a bibliography. Major sections include accessing chemical information via the computer; computer skills used in the chemical industry such as spreadsheets, instrument interfacing, and graphic visualization; computational methods such as ab initio techniques and molecular modeling; and teaching chemistry with computers.

Chapter 2

Applied Sciences Related to Chemistry

This chapter discusses sources that have information at the borders of chemistry: physics, biology—any area that might have answers to questions related to chemistry, and may even have some chemical information. These resources are not limited to chemistry but are valuable nonetheless. Included are resources on meetings and book reviewing media.

GUIDES TO THE LITERATURE

2.001

Guide to reference books: covering materials from 1985-1990. Supplement to the 10th edition, ed. by Robert Balay; special editorial advisor, Eugene P. Sheehy. Chicago; London: American Library Association, 1992. 613 p. 0838905889 Z1035.1 011.02 92-6643

Covers material from 1985 to 1990. Is not superseded or overlapped by the following resource.

2.002

Guide to reference books, 11th ed. Ed. by Robert Balay; assoc. ed., Vee Friesner Carrington; with special editorial assistance by Murray S. Martin. Chicago: American Library Association, 1996. 2,020 p. 0838906699 Z1035.1 011.02 95-26322

Includes bibliographical references and index. Sections: General reference works; Humanities; Social and behavioral sciences; History and area studies; Science, technology, and medicine. (Does not include material from the citation above.)

2.003

Hurt, Charlie Deuel. *Information sources in science and technology*. 3rd ed. Englewood, CO: Libraries Unlimited, 1998. 346 p. (Library and information science text series) 1563085283; 1563085313 pbk Z7401 016.5 98-19547

The author of this bibliographic guide has provided new and notable changes in this third edition while maintaining format, coverage, and the wealth of valuable resources. The principal change is the deletion of titles published before this decade, except those that made "substantive contributions to the field or [were] classic[s] that simply could not be overlooked." Another significant change is the replacement of "Electronic Resources" with "Web Sites." The five major sections are "Multidisciplinary Sources of Information," "Biological Science," "Physical Sciences and Mathematics," "Engineering," and "Health and Veterinary Sciences." Each section lists forms of publication, such as dictionaries, abstracts and indexes, handbooks, encyclopedias, directories. Each entry gives full bibliographic information and critical annotations.

2.004

Walford's guide to reference material: v. 1, Science and technology, ed. by Marilyn Mullay and Priscilla Schlicke. 7th ed. London: The Library Association, 1996- . 967 p. 1856041654 Z1035.1 011.02 97-146982

Annotated, classified list of standard reference books and bibliographies. Viewpoint is that of the U.K.

ABSTRACTING AND INDEXING SERVICES

2.005

Wilson applied science and technology abstracts: CD-ROM. NY: H.W. Wilson, March 1993- . Monthly. ISSN 1076-7185 Z7913 605 96-647356

(From 1913 to 1957, printed *Industrial Arts Index*; from 1958-1993, *Applied Science and Technology Index*.) Cumulative subject index and abstracts to about 335 English-language periodicals in aeronautics and space science, automation, chemistry, construction, earth sciences, electricity, electronics, engineering, industrial and mechanical arts, materials, mathematics, metallurgy, physics, telecommunications, transportation, and related subjects. No author entries except a separate listing for authors of books reviewed. Available in an online format through Wilsonline.

2.006

Wilson general science abstracts: CD-ROM. NY: H.W. Wilson, 1994- . Monthly. ISSN 1076-7177 Q1 500 97-644595

Print indexing service began with v. 1, July 1978. Cumulative abstracts to more than 100 English-language general science periodicals not completely covered by other abstracts and indexes. Author listing of citations to book reviews. Also available online via Wilsonline.

GENERAL ENCYCLOPEDIAS

2.007

Encyclopedia of physical science and technology, ed. by Robert A. Meyers. 3rd ed. San Diego, CA: Academic Press, c2001. 18 v. 0122274105 Q123 503 2001-090661

With more than 700 articles, 10,000 illustrations and tables, and 6,500 bibliographic entries, this encyclopedia covers scientific endeavor in the fields of physical science, mathematics, engineering, and technology. Articles are written by experts and treatment is exhaustive. Each article contains the affiliation of the author, subtitles for the sections of the article, a helpful glossary of terms for the topic, the text of the article, a bibliography, and a list of tables and figures. Computer-readable version is available (see next entry).

[Academic Press] Encyclopedia of physical science and technology [computer file]. Mountain View, CA: Knight-Ridder Information, Inc.; San Diego, CA: Academic Press [producer], 1995. 2 CD-ROMs. At head of title: KR Information OnDisc. 0120002000 Q121

Computer-readable version of the printed service above. Includes 18 volumes of *Encyclopedia of Physical Science and Technology* from Academic Press, 2nd ed., 1995.

2.008

The Facts on File encyclopedia of science. NY: Facts on File, 1999. 2 v., 840 p. (The Facts on File science library) 0816040087 (set); 0816040060 (v. 1); 0816040079 (v. 2) Q123 503 98-53201

Combining comprehensiveness with affordability, this set offers clear, concise entries arranged alphabetically by common rather than scientific name, with numerous cross-references. Diagrams, photographs, and tables supplement the text. Biographical sketches of noted scientists are included, along with articles on such topics as antibiotic resistance, neural networks, cloning, and energy resources.

Internet sites are included with some entries, and the encyclopedia ends with useful appendixes.

2.009

McGraw-Hill concise encyclopedia of science & technology, ed. by Sybil P. Parker. NY: McGraw-Hill, 1998. 4th ed. 2,318 p. 0070526591 Q121 503 98-3875

The 74 consulting editors for this new edition number five more than for the third edition, but there are 100 fewer entries (7,900 in the third edition, 7,800 in the fourth). Both editions include graphs, drawings, diagrams, photographs, signed entries, and bibliographies; the fourth edition updates the bibliographies. Although much the same information is found in the appendixes of both editions, the following appear only in the third: the periodic table, semiconductor symbols and abbreviations, schematic electronic symbols, and classification of living organisms. The following appear only in the fourth: units of temperature in measurement systems and telescopes. Because of their differences, the editions complement one another.

2.010

McGraw-Hill encyclopedia of science & technology. 9th ed. NY: McGraw-Hill, 2002. 20 v. 1st ed., 1960. 0079136656 (set) Q121 503 2001-57910

A comprehensive encyclopedia covering all branches of science and technology. Introductory article provides a broad survey of each branch of science, and separate articles cover the main subdivisions and more specific aspects. Bibliographies follow most longer articles; articles are signed. Kept up-to-date between editions by the *McGraw-Hill Yearbook of Science and Technology*, 1962- . The electronic version, *Access Science*, is available on the Internet at <http://www.accessscience.com/>.

2.011

Van Nostrand's scientific encyclopaedia. 9th ed. Glenn D. Considine, ed. NY; London: Van Nostrand Reinhold, 2002. 2 v., 3,936 p. 0471332305 (set) 1 CD-ROM Q121 503

Includes photographs, diagrams, charts, etc. Approx. 200 contributors. Single, alphabetical arrangement of terms used in science and technology. Entries vary in length. Longer entries have bibliographies and some have Web sites.

2.012

Van Nostrand's scientific encyclopaedia [electronic resource]; ed. by Douglas M. Considine and Glenn D.

Considine. Rev. 8th ed. on CD-ROM. NY: Wiley, 1999. 1 CD-ROM 0471293237 Q121 503 2001-561042

Covers major scientific topics including animal life, biosciences, chemistry, earth and atmospheric sciences, energy sources and power technology, mathematics and information sciences, medicine, anatomy and physiology, physics, plant sciences, space and planetary sciences. More than 7,000 articles written by more than 250 experts.

2.013

Volti, Rudi. *The Facts on File encyclopedia of science, technology, and society*. NY: Facts on File, 1999. 3 v. (Facts on File science library) 0816031231 (set); 0816034591 (v. 1); 0816034605 (v. 2); 0816034613 (v. 3) Q121 503 98-39014

Focusing on the societal context and impact of advances in science and technology, this is the first encyclopedia that offers technical detail together with "historical, cultural, economic and sociological aspects" of each topic. A useful initial reference for general readers that will sensitize students to the broader effects and the philosophical and political underpinnings of science and technology. About 900 entries, (from 500 words to two or three double-column entries) are written by the editor, others by contributors. Few bibliographies. Besides inventions, products, processes, etc., concepts such as "normal accidents" and "spinoffs" are well covered. Reprinting the entire index in each volume is very useful, especially for cross-references.

DICTIONARIES, INCLUDING FOREIGN LANGUAGE

2.014

De Vries, Louis. *French-English science and technology dictionary*. 4th ed. rev. and enlarged by Stanley Hochman. NY: McGraw-Hill, 1976. 683 p. 0070166293 Q123 503 75-45091

1st-3rd ed. (1940-1962) published under the title: *French-English Science Dictionary for Students in Agricultural, Biological, and Physical Sciences*.

2.015

De Vries, Louis, and Theo M. Herrmann. *German-English technical and engineering dictionary*. 2nd ed., completely rev. and enlarged. NY: McGraw-Hill, 1965. 1,178 p. T9 603 65-23218

2.016

Dorian, Angelo Francis. *Dictionary of science and technology: English-French*. NY and Amsterdam: Elsevier, 1979. 1,586 p. 044418296 Q123 503 79-18507

Includes more than 150,000 terms from some 100 fields. Entries indicate scientific field to which term belongs and French equivalent.

2.017

Dorian, Angelo Francis. *Dictionary of science and technology: French-English*. NY and Amsterdam: Elsevier, 1980. 1,086 p. 044441911x Q123 503 80-18312

Companion to English-French work above.

2.018

Dorian, Angelo Francis. *Dictionary of science and technology: English-German*. 2nd rev. ed. NY and Amsterdam: Elsevier, 1978. 1,401 p. 0444416498 Q123 503 77-16111

2.019

Dorian, Angelo Francis. *Dictionary of science and technology: German-English*. 2nd rev. ed. NY and Amsterdam: Elsevier Scientific, 1981. 1,119 p. 0444419977 Q123 503 81-208536

2.020

Erb, Uwe and Harald Keller. *Scientific and technical acronyms, symbols, and abbreviations*. NY: Wiley-Interscience, 2001. 2,100 p. 0471388025 Q179 501 2001-17598

A monumental 2,100-page reference source, comprehensively covering 72 scientific and technical fields, including not only aeronautics, anatomy, botany, geology, mathematics, nuclear science, and physics, but also fields whose abbreviations and acronyms are more difficult to verify (crystallography, forestry, housing, nanotechnology, petrology, textile technology, and others). Entries (more than 200,000) are arranged in a single alphabet; supplementary information appears in either parentheses or brackets.

2.021

McGraw-Hill dictionary of scientific and technical terms. 5th ed. Sybil P. Parker, ed. in chief. NY: McGraw-Hill, 1994. 2,194 p. Appendix. Q123 503 93-34772

Includes 100,100 terms with 117,500 definitions; 102

scientific and technical fields represented; pronunciation guide. Contains disk (IBM-PC version).

2.022

The new Penguin dictionary of science; editor, M.J. Clugston; advisors, N.J. Lord ... [et al.]. London; NY: Penguin Books, 1998. 845 p. (Penguin reference) 0140512713 pbk Q123 503 99-179482

An entirely new dictionary to replace the Penguin publication *Dictionary of Science* by E.B. Uvarov (1942), in print for 50 years. Definitions span chemistry, physics, and mathematics, with fewer from astronomy and biology. Simple, precise definitions, most with cross-references. More than 500 diagrams; appendixes with SI units, decimal multiples, fundamental constants, derivations and integrals, solar system, periodic table, amino acids, and basic biological classifications.

2.023

Malyavska, Greta, and Natalia Shveyeva. *Russian-English dictionary of scientific and engineering terms.* NY: Begell House, 1999. 1,222 p. 1567001289 Q123 503 99-13566

This dictionary from a respected scientific publisher will be useful for those seeking to read or translate technical and scientific texts from Russian into English. It provides English equivalents for technical terms ("parabolicity," "para-hydrogen"), nontechnical words that occur often in the literature ("pair," "section"), and those ordinary phrases ("proved to be the case," "when this happens") that constitute the glue of any language. Each word or phrase is illustrated in the context of a Russian passage, followed by an English rendition.

HANDBOOKS AND DATA COMPILATIONS

2.024

Hall, Carl W. *Laws and models: science, engineering, and technology.* Boca Raton, FL: CRC Press, 2000. 524 p. 0849320186 Q40 500 99-29327

Includes more than 1,500 of the most common or important laws and models (principles, theorems, canons, axioms), alphabetically arranged, with year of promulgation or discovery, brief definition, mathematical expressions, keywords, full name of discoverer, profession, and references to the literature.

2.025

Horvath, Ari L. *Conversion tables of units in science & engineering.* NY: Macmillan, 1986. 147 p. 0333408578 TA151 502 86-5026

Offers values and definitions, and corresponding conversion factors for SI units. Historically important units and systems are found in a separate index.

2.026

International critical tables. E.W. Washburn, ed. NY: McGraw-Hill Book Co., 1926-1930. 7 v. data, index v. Q199

Data compilation is not current beyond 1930. Data for pure chemical and physical material, both single substances and compounds, natural products, certain biological systems, and the solar system, are presented in tabular form, with full references and extensive indexing. Text in English, French, German, and Italian. Literature citations appear at end of each section. Brought up to date by National Standard Reference Data System in a series of monographs. Now available online from Knovel.com.

DIRECTORIES

2.027

America's corporate families: The billion dollar directory. Parsippany, NJ: Dun & Bradstreet, Inc., annual. 3 v. 1562036475 (v. 1-2); 1562036491 (v. 3) HG4057 338.8 82-643626

Lists U.S. corporations and subsidiaries.

2.028

Encyclopaedia of associations, 3rd ed.- . Detroit: Gale Research Company, 1961- . Annual, 1975- . ISSN 0071-0202 AS22 (HD2425) 060 76-46129

Published in 1959 under the title *Encyclopedia of American Associations*. V. 1 contains national organizations of the U.S.; v. 2 is a geographic and executive index. The main body of the work is organized by type of association; there is a name index. Entries give date of establishment, type of activities, name of publication(s), chief executive officer, and address and telephone number of headquarters.

2.029

Million dollar directory: America's leading public

and private companies. Parsippany, NJ: Dun's Marketing Services, 1959-1996. 38 v. ISSN 0070-7619 HC102 338.7 59-3033

Arrangement is as follows: names, alphabetically arranged; cross-references by geography; cross-references by industry.

2.030

Moody's industrial manual. NY: Moody's Investors Service, 1954- . Annual. ISSN 0545-0217 HG4961 332.6 56-14721

"Covering New York, American & regional stock exchanges & international companies." Volumes from 1972- issued in parts; updated semiweekly by *Moody's Industrial News Reports*. Corresponding CD-ROM publication: *Moody's Company Data*. Also available via the Internet.

2.031

Standard and Poor's register of corporations, directors, and executives, United States and Canada. NY: Standard & Poor's Corp., 1973- . Annual. HG4057 87-17041

Title varies slightly. Some editions in multiple volumes. Kept up to date between editions by three cumulative supplements each year. Geographical index.

2.032

Thomas register of American manufacturers. NY: Thomas Publishing Company, 1906- . Annual. ISSN 0362-7721 T12 338.7 06-43937

Lists, by product, more than 152,000 U.S. and 9,000 Canadian manufacturers, vendors, etc. of all types of manufactured goods and services. Includes trade names and address lists, with telephone, telex, 800, and fax numbers. Multiple volumes since 1944. Available online gratis.

2.033

Who's who in science and engineering 1998-1999. 4th ed. New Providence, NJ: Marquis Who's Who, 1997. 1,638 p. 0837957567 (Classic ed.); 0837957575 (Deluxe ed.) Q141 79-139215

Each biography contains full name, occupation, date and place of birth, parents, spouse, children, education, career summary, writings and creative works, civic and political activity, military record, awards, association memberships, clubs, religion, and home and office addresses. More than 25,000 scientists are profiled.

2.034

The World of learning. London: Europa Publications, 1948- . Annual. (1st ed., 1947, published by Allen & Unwin, London.) ISSN 0084-2117 AS2 060.25 47-30172

This European publication lists, by country, the world's colleges and universities (including faculty listings where available), academies, institutes, museums, international organizations, UNESCO activities, and learned societies.

FORMULARIES ("RECIPE BOOKS")

2.035

The chemical formulary, ed. by H. Bennett. Brooklyn, NY: Chemical Publ. Co., v. 1, 1933 to v. 15, 1970; 3 v. Irregular. TP151 660.831 89-29408

V. 1-6, 1933-1943; v. 1-10, 1933-1957; v. 1-15, 1933-1970. "A collection of valuable, timely, practical commercial formulae and recipes for making thousands of products in many fields of industry." Covers all types of recipes, from brushless shaving cream to aquarium cement, hand cream, and rat and ant poison.

2.036

Henley's twentieth century book of formulas, processes, and trade secrets; a valuable reference book for the home, factory, office, laboratory and the workshop; ... , ed. by Gardner D. Hiscox. New rev. and enl. ed. by T. O'Conor Sloane; rev. 1956 by Harry E. Eisenson. Cornwells Heights, PA: Publishers Agency, 1981. 867 p. T49 603 76-18904

Previously published under titles: *Henley's Twentieth Century Book of Recipes, Formulas and Processes*; *Henley's Twentieth Century Formulas, Recipes and Processes*; *Fortunes in Formulas*.

BOOK SELECTION MEDIA

2.037

AAAS science book list, 1978-1986, compiled and ed. by Kathryn Wolff, Susan M. O'Connell, and Valerie J.

Montenegro. Washington, DC: American Association for the Advancement of Science, 1986. 568 p. 0871683156 Q181 500 86-26519

Supplements the *AAAS Science Book List* (3rd ed., 1970) and the *AAAS Science Book List Supplement* (1978). Includes reviews appearing in *Science Books & Films*, 1978-1986.

2.038

Choice: Current reviews for college libraries. Chicago: American Library Association, Mar. 1964- . Monthly (bimonthly July/Aug). ISSN 0009-4978 Z1035 028 64-9413

Eleven issues/year; about 600 reviews of books and electronic products in each issue, classed loosely from reference through all subject fields. Has science and technology areas, including chemistry. Annually, has a special listing of books categorized as "Outstanding Academic Titles," reflecting their reviewers' opinions and selected as such by the editors. Since 1997, publishes a supplement in August with Internet site reviews. Reviews since 1989 available via subscription on the Internet.

2.039

Science books & films. v. 11:1- , 1975- . Washington, DC: American Association for the Advancement of Science. 5/yr. (Title varies: *AAAS science books & films*). ISSN 0098-342x Z7403 016.5 75-645493

Continues *Science Books: A Quarterly Review* (v. 1-10, 1965-1975). Annotated, classified listing of new books, films, filmstrips, and videocassettes in the pure and applied sciences. Provides a useful selection guide for libraries serving students from elementary level through first two years of college.

BIOGRAPHICAL DIRECTORIES

2.040

American men & women of science: a biographical directory of today's leaders in physical, biological, and related sciences. 21st ed. Detroit, MI: Thomson/ Gale, 2003. 8 v. 0787665231 (set) Q141 509.22

Now in its 96th year; contains full biographical information on nearly 130,000 living scientists in the U.S. and Canada. Alphabetically arranged by name; v. 8 is a discipline index based on the National Science Foundation's Taxonomy of

Degree and Employment Specialties. Began in 1906 as *American Men of Science*.

2.041

Bailey, Martha J. *American women in science, 1950 to the present: a biographical dictionary*. Santa Barbara, CA: ABC-CLIO, 1998. 455 p. 0874369215 Q141 500 98-22433

In this volume, the time focus moves from the first half of the 20th century to the second, with liberal inclusion of tangential fields (e.g., anthropology, economics, psychology, sociology). The physical, biological, medical, environmental, mathematical, and applied sciences are well represented. An introductory essay on the status and progress of women in science since WW II precedes the 304 biographies of women in 78 professions. Entries are one to two pages in length; 52 include photographs. In the text, scientific terms receive parenthetical definitions.

2.042

Biographical encyclopedia of scientists, ed. by Richard Olson with Roger Smith. NY: Marshall Cavendish, 1998. 5 v., 1,460 p. 0761470646 (set) Q141 509 97-23877

(This set should not be confused with another by the same title published in 1994.) Subjects have been selected primarily to display the broad spectrum of disciplines, nationalities, races, and genders that characterized scientific endeavor throughout history to the present day. Signed biographies; subject's area of achievement is followed by a statement of that individual's contribution to science; time line listing birth and death dates, major awards and honors, and milestones in the scientist's education, research, employment, and private life; short biographical essay; short article and bibliography on the scientist's specialty; brief bibliographies of works by and about the individual. There are excellent photographs or woodcuts for each individual. Key to pronunciation; glossary; general time line of scientific events; area-of-specialty list; comprehensive index.

2.043

Dictionary of scientific biography. Charles Coulston Gillispie, ed. in chief. NY: Scribner, 1970-1980. 16 v. 0684101149 (v. 3) Q141 509.2 69-18090

V. 1-14, A-Z; v. 15: Supplement 1; v. 16: Index. Includes scientists from all periods of history (excluding living persons) representing all fields of science.

2.044

Dictionary of scientific biography. Charles Coulston Gillispie, ed. in chief. NY: Scribner, 1981-<1990 >. v. <1-18; in 10 > 0684169622 (set) Q141 509.2 80-27830

Published under the auspices of the American Council of Learned Societies. Beginning with v. 15, issued as supplements. V. 17-<18 >, Frederic L. Holmes, editor in chief. Biographies are short, some a half-page long, some a page and a half. Bibliographies included. Data include birth and death dates, birthplaces, and fields of scholarship.

2.045

Howsam, Leslie. *Scientists since 1660: a bibliography of biographies*. Aldershot, England; Brookfield, VT: Ashgate, 1997. 150 p. 1859280358 Z7401 016.51 96-41789

Selected from the *Dictionary of Scientific Biography* (1981-) are names of 565 men and women in science whose careers began since 1660 and whose lives and work have been the subject of one or more book-length biographies or an autobiography. The author searched out 1,106 titles dealing with these individuals, excluding short or cumulative biographies; for each scientist, pertinent works are listed in chronological order, giving complete bibliographic data plus a brief annotation derived from the book's preface or introduction.

2.046

Notable black American scientists, ed. by Kristine Krapp. Detroit: Gale Research, 1999. 349 p. 0787627895 Q141 509.2 98-36338

The 52 contributors to this compilation of 254 bibliographic profiles emphasize the achievements of black scientists and physicians, men and women, from Colonial times to the present, in the territory that is now the US. Each entry begins with basic information about each subject—name, year of birth and death (if deceased), and specialty. A biographical essay follows, ranging from about 400 to 2,000 words and covering the subject's life and professional accomplishments. Most essays are followed by "Selected Writings by the Scientist" and "Further Reading." A time line (1619-1995) includes scientific achievements of the subjects and significant events in African American history. Indexes by gender, fields of specialization, and subjects.

2.047

Notable twentieth-century scientists, ed. by Emily J.

McMurray. Detroit: Gale (ITP), 1995. 4 v. 0810391813 (set) Q141 509.22 94-5263

Entry list in alphabetical order; "chronology of scientific advancement," 1895-1993; biographies cover living and deceased. Portraits; bibliographies, biographies cover one to three pages; about 1,000 biographies.

2.048

Notable twentieth-century scientists. Supplement, ed. by Kristine M. Krapp. Detroit: Gale, 1998. 617 p. 0787627666 Q141 509.2 98-14016

Gale's parent collection (*Notable Twentieth-Century Scientists*, 4 v.) consists of biographical sketches of nearly 1,300 scientists, all of whom lived during part of the 20th century, and tried to include as many women, US ethnic minorities, and scientists from outside North America and Western Europe as possible. This supplement includes 250 new biographies and 65 updates. The "Selected Biographical Sources," "Field of Specialization Index," "Gender Index," "Nationality/Ethnicity Index," and "Subject Index" all have been updated and now include the original volumes and this supplement.

2.049

Notable women in the physical sciences: a biographical dictionary, ed. by Benjamin F. Shearer and Barbara S. Shearer. Westport, CT: Greenwood Press, 1997. 479 p. 0313293031 Q141 500.2 96-9024

Featuring biographical essays on 96 world and US women scientists, this volume includes women who made a significant contribution to the physical sciences from antiquity to the present, though the emphasis is on 20th-century women. Many are still living, and the essays include quotes from recent interviews. Disciplines include astronomy, astrophysics, biochemistry, chemistry, and physics. The essays average five pages in length, and describe obstacles encountered and achievements experienced by each scientist. Entries provide a chronology, a descriptive essay, and a bibliography. Includes 47 photographs. This is a companion volume to *Notable Women in the Life Sciences*, also edited by the Shearers.

2.050

Smith, Roger. *Biographies of scientists: an annotated bibliography*. Lanham, MD: Scarecrow Press; Pasadena, CA: Salem Press, 1998. 291 p. 0810833840 Z7404 016.5092 98-5954

The goal of this work is to guide general readers to 736

volumes that might offer full-length biographies of scientists. The books are grouped by multidisciplinary sources, astronomy and cosmology, chemistry, earth sciences, life sciences, mathematics, medical sciences, physics, and related fields. Each group is divided by collections, and by biographies and autobiographies, listed by author. Bibliographical data and short reviews are presented for each volume. Because of interdisciplinary interests and the frequent listing of more than one book about a scientist, the name and subject indexes are very useful. The book also provides a list of Web sites about scientists.

2.051

Warren, Wini. *Black women scientists in the United States*. Bloomington: Indiana University Press, 1999. 366 p. 0253336031 Q141 500 99-40264

Biographical sketches, alphabetically arranged, of more than 100 black women scientists, living and deceased. Disciplines represented include anatomy, anthropology, astronautics and space science, biochemistry, biology, chemistry, geology, marine biology, mathematics, medicine, nutrition, pharmacology, physics, psychology, and zoology. Indexes of personal names, and disciplines.

2.052

Yount, Lisa. *A to Z of women in science and math*. NY: Facts on File, 1999. 254 p. 0816037973 Q141 509.2 98-46093

This biographical dictionary profiles 150 women whose research has made direct contributions to science. Both present-day scientists and those from earlier periods from all countries are included. Entries are brief (one to two pages in length), are arranged alphabetically, and include bibliographies. Biographical sketches contain family background, education, personal obstacles overcome, and details about scientific work. Numerous indexes list the women by time period, field of study, country of birth, and country of scientific activity. Fields covered range from astronomy and mathematics to physics and zoology.

HISTORY OF SCIENCE

2.053

A Century of Nobel Prize recipients—chemistry, physics, and medicine, ed. by Francis Leroy. NY: Marcel Dekker, 2003. 300 p. 0824708768 Q141 509.2 2002-44452

Prize recipients from 1901 through 2001 are profiled; there is a bibliography, index of Prizewinners' names, and a subject index.

2.054

Encyclopaedia of the history of science, technology, and medicine in non-Western cultures, ed. by Helaine Selin. Dordrecht; Boston: Kluwer Academic, 1997. 1,117 p. 0792340663 Q124.8 509 96-36625

Early Western culture relied much on the science of other cultures, but as academia became more Eurocentric, much of the science in non-Western cultures was undervalued or lost. Some 600 signed essays by eminent scientists and historians of non-Western science are a landmark in the study of the history and philosophy of science, technology, and medicine. The articles consist of brief biographies, some long philosophical articles, and general articles about the comparison and history of sciences among various cultures. Some articles discuss the relationship between colonialism and science, environment and nature, maps and map making, and magic and science. Each article includes a short list of references.

2.055

Isis cumulative bibliography; A bibliography of the history of science formed from Isis critical bibliographies 1-90, 1913-1965, ed. by Magda Whitrow. London: Mansell, with the History of Science Society, 1971-<1984 >. 6 v. 0720101832 Z7405 016.509 72-186272

V. 1, pt. 1: Personalities, A-J; v. 2, pt. 1: Personalities, K-Z. pt 2: Institutions; v. 3: Subjects; v. 4: Civilizations and periods, prehistory to Middle Ages; v. 5: Civilizations and periods, 15th to 19th centuries; v. 6: Author index.

2.056

Isis cumulative bibliography; A bibliography of the history of science formed from Isis critical bibliographies 91-100 indexing literature published from 1965 through 1974, ed. by John Neu. London: Mansell, with the History of Science Society, 1980-1985. v. 1-2. (In progress).

Continues the series above.

2.057

Isis cumulative bibliography, 1986-95: A bibliography

of the history of science formed from the annual *Isis* current bibliographies, ed. by John Neu. Canton, MA: Published for the History of Science Society by Science History Publications/USA, 1997. 4 v. 0881351318 (v. 1); 0881351326 (v. 2); 0881351334 (v. 3); 0881351342 (v. 4) Z7405.H6 016.509 97-18452

Isis, "dedicated to the history of science," was started by George Sarton in 1913. Its annual bibliographies have both stimulated and documented international research in this field. The *Isis Cumulative Bibliography, 1986-95* is the third supplement to the original cumulation, edited by Magda Whitrow, which covered *Isis* bibliographies published from 1913 through 1965. The division of this cumulation into three parts—persons and institutions, subjects, and chronological periods—as well as the subject classification scheme, conforms to the earlier *Cumulative Bibliographies*, and a separate section is again devoted to book reviews.

annual cumulative index. ISSN 0032-9568 Z5063 016.6291 65-9856

Indexes all proceedings in print, rather than information about when conferences are held.

2.061

World meetings: United States and Canada. NY: Macmillan Information, v. 1- , Sept. 1963- . ISSN 0043-8693 Q11 506 75-649498

Title in 1963 was *Technical Meetings Index*. Contains meeting calendar, papers, and personnel involved in meetings.

PROCEEDINGS AND MEETING CALENDARS

2.058

Eventline. The Netherlands: Excerpta Medica Medical Communications, 1989- . Available via STN and other sources.

This directory covers past, current, and future conferences, exhibitions, trade fairs, symposia, meetings, and other events worldwide, emphasizing medicine, biotechnology, and the sciences. Searchable portions include meeting titles, dates, locations, organizers, and more than 700 subject classifications.

2.059

ISI Proceedings. Science & Technology edition. Philadelphia: Thomson/ISI, 2001- . Part of ISI Web of Knowledge (online).

A multidisciplinary product that offers access to bibliographic information and author abstracts of papers from conferences, symposia, seminars, colloquia, workshops, and conventions. Cross-referenced with other ISI electronic products. About two million papers from about 60,000 conferences since 1990 are available.

2.060

Proceedings in print. Arlington, MA: Proceedings in Print, Inc., v. 1- , 1964- . Five numbers/year, with separate

Chapter 3

Analytical Chemistry

As promised in Chapter 1, this is a journey through the chemical literature with this book as a guide, highlighting important landmarks along the way. It includes classic works, such as landmark treatises, old, well-known handbooks, and some new additions to this very important group of chemical information sources.

Readers may wonder why certain chemical techniques are not found in this section. Placement of these techniques depends on how the technique is treated; for instance, electrochemistry, which may in reality be a physical chemical entity, is placed with physical chemistry. If such a technique states that it is an analytical technique, then it will be found in this chapter. Analytical techniques that are general, such as a spectroscopic work treating more than one branch of chemistry, will be found here; single-subject techniques volumes are included with their respective subject areas.

Treats the basic and applied aspects of analytical chemistry. Long articles discuss theoretical topics, particular techniques of analysis, and analytical methods for major groups of compounds, for individual elements, for particularly important specific compounds, and for many commercial products. Articles are thorough but not detailed. Each article has five to ten references, citing mainly books, treatises, and review articles, only occasionally research papers. Good illustrations, some in color; frequent, relevant tables. Index volume.

GENERAL HANDBOOKS AND DATA COMPILATIONS

3.003

Analytical instrumentation handbook, ed. by Galen Wood Ewing. 2nd ed., rev. and expanded. NY: M. Dekker, 1997. 1,453 p. 0824794559 QD79.I5 543 96-51434

Each technique is separately authored. Contents: The laboratory use of computers; Laboratory balances; Organic elemental analysis; Continuous-flow analyzers; Atomic emission spectroscopy; Atomic absorption and flame emission spectrometry; Ultraviolet, visible, and near-infrared spectrophotometers; Molecular fluorescence and phosphorescence; Vibrational spectroscopy: instrumentation for infrared and Raman spectroscopy; X-ray methods; Photoacoustic spectroscopy; Techniques of chiroptical spectroscopy; Nuclear magnetic resonance; Electron paramagnetic resonance; X-ray photoelectron and auger electron spectroscopy; Mass spectrometry; Thermoanalytical instrumentation and applications; Potentiometry: pH and ion-selective electrodes; Voltammetry; Instrumentation for stripping analysis; Measurement of electrolytic conductance; Basic instrumentation for chromatography: high-performance liquid chromatography; Modern gas chromatographic instrumentation; Supercritical fluid chromatography instrumentation; Capillary electrophoresis.

ENCYCLOPEDIAS

3.001

Encyclopedia of analytical chemistry: applications, theory, and instrumentation; editor in chief, Robert A. Meyers. New York: Wiley, 2000. 15 v. 0471976709 (set) QD71.5 543 00-42282

A comprehensive reference covering theory and instrumentation through applications and techniques. More than 600 articles, alphabetically arranged by topic. Many color illustrations; extensive bibliographies; expert authorship of articles. All fields covered: analytical, organic, physical, polymer, inorganic, biomedical, environmental, pharmaceutical, industrial, petroleum, forensics, and food science.

3.002

Encyclopedia of analytical science, ed. by Alan Townshend. London; San Diego, CA: Academic Press, 1995. 10v., 6,059 p. 0122267001 (set) QD71.5 543 94-24914

3.004

Association of Official Analytical Chemists. *Handbook for AOAC members.* 6th ed. Arlington, VA:

Association of Official Analytical Chemists, 1989. 64 p. 0935584412 QD71

A useful reference for all AOAC members involved in methods research, publication, meetings, committees, regional sections, and liaison with other scientific organizations.

3.005

Bruno, Thomas J., and Paris D.N. Svoronos. *CRC handbook of basic tables for chemical analysis.* Boca Raton, FL: CRC Press, 1989. 517 p. 0849339359 QD78 543 88-7595

Covers both instrumental methods and wet chemical techniques. Features laboratory safety, hazardous chemicals handling, and qualitative tests for organic and inorganic species. Treats gas, high-performance liquid, thin layer, and supercritical fluid chromatography; ultraviolet, infrared, and atomic absorption spectrometry; and nuclear magnetic resonance and mass spectroscopy.

3.006

Dean, John Aurie. *Analytical chemistry handbook.* NY: McGraw-Hill, 1995. Separately paged sections. 0070161976 QD78 543 94-47607

Twenty-three chapters treat preliminary operations of analysis; preliminary separation analysis; gravimetric and volumetric analysis; chromatographic methods; electronic absorption and luminescence spectroscopy; infrared and Raman spectroscopy; atomic spectroscopy; optical activity and rotary dispersion; refractometry; X-ray methods; radiochemical methods; nuclear magnetic resonance spectroscopy and electron spin resonance; mass spectroscopy; electroanalytical methods; thermal analysis; magnetic susceptibility; organic elemental analysis; detection and determination of functional groups in organic compounds; determination of water; statistics in chemical analysis; geological and inorganic materials; water analysis; and general information.

3.007

Handbook of analytical chemistry. Louis Meites, ed. 1st ed. NY: McGraw-Hill, [1963]. Separately paged sections; 1,806 p. QD71 543.082 61-15915

Separation, analysis, identification of substances. Over 100 contributors and over 60 different techniques covered.

3.008

Handbook of analytical techniques, ed. by Helmut Gèunzler and Alex Williams. Weinheim; Cambridge: Wiley-VCH, 2001. 2 v., 1,200 p. 3527301658 543

Concise, one-stop reference for spectroscopic, chromatographic, and electrochemical techniques, including chemical and biochemical sensors, thermal analysis, and bioanalytical, nuclear, and radiochemical techniques.

3.009

Handbook of instrumental techniques for analytical chemistry, ed. by Frank A. Settle. Upper Saddle River, NJ: Prentice Hall PTR, 1997. 995 p.; CD-ROM 0131773380 QD79.I5 543 97-10618

This book combines analytical instrumentation material with analytical applications and a guide to available instruments and their procurement. Sampling techniques, general analytical considerations, separation methods, spectroscopy, mass spectrometry, electroanalytical techniques, microscopy, surface analysis, and polymer examination are included; the CD-ROM contains software that links analysis techniques to references within the handbook.

3.010

Reagent chemicals: American Chemical Society Specifications, Official from January 1, 2000. 9th ed. New York: ACS, 2000. 752 p. 0841236712 QD77 543 99-38870

Includes index. Covers chemical tests, reagents, and indicators. Alphabetic by reagent name; alternate name, formula, molecular weight, CASRN, requirements; use in which tests and how done. Also available via the Internet.

ANALYTICAL METHODS

Standard Methods

3.011

Compilation of EPA's sampling and analysis methods, ed. by Lawrence H. Keith. 2nd ed. Boca Raton, FL: CRC/Lewis Publishers, 1996. 1,696 p. 1566701708 TD193 628.1 96-25999

This new edition contains nearly 1,200 analyte/method summaries with descriptions, required instrumentation, interferences, sampling containers, preservation techniques, maximum holding times, detection levels, accuracy, precision, quality control, EPA references, and (where available) EPA contacts and telephone numbers.

3.012

Official methods of analysis of the AOAC International. 16th ed. Arlington, VA: AOAC International, 1995- . 2 v., looseleaf. Annual. ISSN 1080-0344 S587 630 96-656222

Includes more than 1,800 collaboratively tested methods for analyzing chemical and biological substances. Includes bibliographical references. Supplements issued between major revisions. Contents: v. 1: Agricultural chemicals; contaminants; drugs; v. 2: Food composition; additives; natural contaminants. Includes definitions of terms, guide to method format, standard solutions and certified reference materials, laboratory safety, index. Also available in electronic format.

3.013

Scholz, Eugen. *Karl Fischer titration: determination of water*. Berlin: Springer-Verlag, 1984. 138 p. 0387137343 QD511 544 84-014165

Provides a new formulation of the KF reaction and describes all methods in full detail, from the classic volumetric titration to modern coulometric water determination, from conventional equipment to the latest reagents. Also covers a broad range of applications in organic and inorganic compounds, in foodstuffs, and in natural and artificial substances. Safety precautions are provided.

3.014

Standard methods of chemical analysis. 6th ed. Previous editions edited by W.W. Scott. Princeton, NJ: Van Nostrand, 1962-1966; reprinted by R.E. Krieger Pub. Co., 1975- . 3 v. in 5 v. 0882752545 QD131 543 74-23465

Compilation of current methods of analysis. V. 1: The elements; Methods of determining each element in natural condition and in synthetic products. Arranged in alphabetical order with logical groupings. V. 2: Industrial and natural products and non-instrumental methods; 2 parts: Natural and manufactured products; Details of procedure and apparatus. V. 3: Instrumental analysis.

3.015

Thurman, E.M., and M.S. Mills. *Solid-phase extraction: principles and practice*. New York: Wiley, 1998. 344 p. (Chemical analysis, v. 147) (A Wiley-Interscience publication) 047161422x QD63.E88 543 97-27535

Chapters outline basic chemistry of solid phase extraction (SPE); applications of SPE in areas such as environmental analysis, drugs and pharmaceuticals, and foods and natural products; and recent advances in SPE technology including automation, solid phase extraction disks, solid phase microextractions, and molecular recognition SPE. Appendix with Internet sites related to solid-phase extraction.

3.016

Watson, C.A. *Official and standardized methods of analysis*, compiled and edited for the Analytical Methods Committee of the Royal Society of Chemistry. 3rd ed. Cambridge: Royal Society of Chemistry, 1994. 778 p. 0851864414 QD75.2 545 94-81670

Summarizes the methods used in analytical chemistry, including animal feeds, cosmetics, essential oils, fertilizers, meat products, vitamins, milk and milk products, pesticide residues, occupationally hazardous materials, oils and fats, soaps, water analysis, brewing, and other industrially useful techniques.

General Methods

3.017

Advances in analytical chemistry and instrumentation. C.N. Reilly, ed. NY: Wiley-Interscience, 1960-1973. v. 1-11. Irregular. ISSN 0065-2148 QD71 543.082 60-13119

Each volume treats a separate topic in various chapters; individual authors each discuss some aspect of volume's subject. Cumulative index in each volume; volumes issued irregularly and not in sequence.

3.018

Budevsky, O. *Foundations of chemical analysis*, translation editors, R.A. Chalmers & M.R. Masson. Chichester, UK: Ellis Horwood; NY: Halsted Press, 1979. 372 p. (Ellis Horwood series in analytical chemistry) 0470266929 QD75.2 543 79-322387

Strong chapters on complexometry and precipitation equilibrium; graphical solutions to complex systems.

3.019

Chemical analysis in complex matrices, ed. by Malcolm R. Smyth. NY: E. Horwood: PTR Prentice Hall, 1992. 295 p. (Ellis Horwood PTR Prentice Hall analytical chemistry series) 0131276719 QD75.2 543 91-42948

Informative overview of sampling and analytical considerations in the micro- and macroanalysis of chemical species within a matrix. Chapters review analysis of drugs in biological fluids, components in brewed beverages, sealants and adhesives, air pollutants, and human and animal foods.

3.020

Fritz, James S. *Analytical solid-phase extraction.* NY: Wiley-VCH, 1999. 209 p. 0471246670 QD63.E88 543 98-43758

Covers all aspects of solid-phase extraction (SPE) including basic principles, historical perspective, materials and equipment, extraction of organics from aqueous samples, ion exchange, metal ion extraction, microscale extractions, and applications. Numerous examples, figures, drawings.

3.021

Khandpur, Raghbir Singh. *Handbook of modern analytical instruments.* Blue Ridge Summit, PA: TAB Books, 1981. 588 p. 0830611509 QD53 543 81-9133

Concise but comprehensive discussion of construction and operation of the most frequently used electrochemical, spectrometric, and chromatographic instruments.

3.022

Kolthoff, I.M., and V.A. Stenger. *Volumetric analysis.* 2nd rev. ed. Tr. by N.H. Furman. NY: Interscience Publishers, 1942-1957. 3 volumes, each with an author and subject index. QD111 545.5 42-14617

V. 1: Theoretical fundamentals; v.2: Titration. Methods: acid-base, precipitation, and complex-formation reactions; v.3: Titration methods: oxidation-reduction reactions, by I.M. Kolthoff and R. Belcher with the cooperation of V.A. Stenger and G. Matsuyama. V. 2 treats apparatus and general principles; acid-base reactions; quantitative precipitation and complex formation reactions.

3.023

Levinson, Ralph. *More modern chemical techniques.* London: Royal Society of Chemistry, 2001. 184 p. 0854049290 QD75.22

Discusses elemental analysis, such as atomic absorption spectrometry and inductively coupled plasma techniques; separations, including electrophoresis; structure determination, such as X-ray diffraction and optical microscopy; and sampling and sample preparation.

3.024

Mueller-Harvey, Irene and R.M. Baker. *Chemical analysis in the laboratory: a basic guide.* Cambridge: Royal Society of Chemistry, 2002. 92 p. 0854046461 543

Discusses sampling, sample collection and preparation, planning laboratory work, weights and measures, digestion and extraction, and determinations.

3.025

Physical methods in modern chemical analysis. Theodore Kuwana, ed. NY: Academic Press, 1978-83. v. 1, 1978; v. 2, 1980; v. 3, 1983. 0124308015 (v. 1) QD75.2 543 77-92242

V. 3: X-ray spectrometry; Transform techniques in chemistry; Electrochemical characterization of chemical systems; Global optimization strategy for gas-chromatographic separations. Separate essays on methods, written by experts, about four to a volume.

3.026

Rouessac, Francis, and Annick Rouessac. *Chemical analysis: modern instrumental methods and techniques.* Translated by Michel Bertrand and Karen Waldron. Chichester; New York: J. Wiley, 2000. 445 p. 0471981370; 0471972614 pbk QD79.I5 543 99-58872

Focus is on instrumental methods only; comprehensive coverage of numerous analytical techniques and instrumentation, and of latest developments including supercritical fluid chromatography and capillary electrophoresis. Covers chromatography and spectral methods as well as radiochemical, potentiometric, voltammetric, and coulometric methods.

3.027

Separation, purification and identification, ed. by Lesley Smart. Cambridge, UK: Royal Society of Chemistry, 2002. 120 p. 2 CD-ROMs (The molecular world) 0854046852 QD75.22 544

Discusses distillation and recrystallization; chromatography; elemental analysis; atomic absorption, mass, infrared, and nuclear magnetic spectrosopy.

3.028

Taylor, Larry R., Richard B. Papp, and Bruce D. Pollard. *Instrumental methods for determining elements.* NY: VCH, 1994. 322 p. 1560810386 QD79.I5 543 93-38305

Very comprehensive and thorough, covering all of the major analytical methods. An excellent chapter treats factors to be considered in choosing an analytical method. The work covers atomic absorption, atomic emission, electrochemical techniques, potentiometric techniques, chromatographic methods, X-ray fluorescence, combustion, and miscellaneous methods, which include capillary electrophoresis and mass spectrometry.

3.029

Thomas, Leslie C., and Gordon J. Chamberlin. *Colorimetric chemical analytical methods.* 9th ed. Rev. by G. Shute. Salisbury, Eng.: Tintometer; NY: distr. by Wiley, 1980. 625 p. 0471276057 QD113 545 79-56635

1st-7th eds. by Tintometer Ltd.; 1st-3rd issued as *A Handbook of Colorimetric Chemical Analytical Methods for Industrial, Research and Clinical Laboratories.* Tests for air, beverages, biological materials, chemicals, dairy products, foods and edible oils, fuels and lubricants, metals and alloys, minerals, ores and rocks, miscellaneous, paper, plants, plastics and polymers, sewage and industrial wastes, soil, and water.

3.030

Walton, Harold F., and Roy D. Rocklin. *Ion exchange in analytical chemistry.* Boca Raton, FL: CRC Press, 1990. 229 p. 0849361990 QD79.C453 543 89-70839

Describes ion exchanging materials and classical separation and trace concentration techniques as well as ion exchange as a basis for high pressure liquid chromatography ("ion chromatography"). Includes fundamental related material such as Donnan equilibrium, exchange kinetics, conductimetry, and potentiometry.

SPECIAL TOPICS

Chromatography

3.031

Capillary electrochromatography, ed. by Keith D. Bartle and Peter Myers. Cambridge, UK: Royal Society of Chemistry, 2001. 149 p. (RSC chromatography monographs) 0854045309 QP519.9.C36 543.0894

Introduces the hybrid separation methodology of capillary electrochromatography (CEC) that combines the advantages of capillary electrophoresis and high-performance liquid chromatography. Capillary electrochromatography refers to packed column electrochromatography and not to micellar electrokinetic capillary chromatography. Contains a useful table of symbols and abbreviations, and each of the well-referenced eight chapters is written by a leading researcher in the field, in a readable and well-illustrated format. Theory, instrumentation, and range of column properties in CEC, are followed by two modifications: use of open tubular columns and mass spectroscopy detection.

3.032

Chromatography: a laboratory handbook of chromatographic and electrophoretic methods, ed. by E. Heftmann. 3rd ed. NY: Van Nostrand Reinhold, 1975. 969 p. 0442232802 QD79 544 75-5804

Chapters by experts in area; covers high-efficiency liquid column chromatography, affinity chromatography, isoelectric focusing, and pore gradient electrophoresis as the new techniques. Covers old established techniques as well.

3.033

Chromatography: fundamentals and applications of chromatography and related differential migration methods, ed. by E. Heftmann. 5th ed. NY: Elsevier, 1992. (Journal of Chromatography library, v. 51A-B). 0444884041 (set) QD79 543 91-35963

Pt. A: Fundamentals and techniques; Pt. B: Applications.

3.034

CRC handbook of chromatography, ed. by Gunter Zweig and Joseph Sherma. Cleveland, CRC Press, 1972. [1st ed.] 2 v. 0878195602 QD117 544 76-163067

V. 1 lists 549 tables with RF values, retention times and volumes, and other data for gas, liquid column, paper, and thin-layer chromatography. Each table lists reference sources for its data. V. 2 contains techniques, reagents, and sample preparation methods. (For later parts/editions, see under CRC Handbook Series in Chromatography in this section.)

3.035

Encyclopedia of chromatography, ed. by Jack Cazes. NY: Marcel Dekker, 2001. 927 p. 0824705114 QD79.C4 543 2001-28927

More than 300 articles by some 180 experts discuss history,

varieties of technique, methodologies, current state of the art, and future perspectives in chromatography. Arrangement is alphabetical; subject and author indexes, including citations.

3.036

Gas chromatography in forensic science, ed. by Ian Tebbett. NY: E. Horwood, 1992. 188 p. (Ellis Horwood series in forensic science) 0133271986 RA1057 614 92-30473

Well-written and thoroughly researched contributed chapters that describe analysis of blood alcohol; analysis of street samples for drugs of abuse; analysis of biological fluids for drugs; and analysis of debris for accelerants and explosives. Contents: Forensic gas chromatography; Gas chromatography and forensic drug analysis; Gas chromatographic applications in forensic toxicology; Analysis of alcohol and other volatiles; Gas chromatography in arson and explosive analysis; Pyrolysis gas chromatography in forensic science.

3.037

HPLC and CE: principles and practice, [ed. by] Andrea Weston and Phyllis R. Brown. San Diego: Academic Press, 1997. 280 p. 0121366405 QD79.C454 543 96-49198

Outlines basic concepts, chemistry, and instrumentation of high-performance liquid chromatography (HPLC) and capillary electrophoresis (CE); compares these methods for solving chemical separation problems. Treats normal and reverse phase separations, capillary zone electrophoresis, micellar electrokinetic capillary chromatography, capillary gel electrophoresis, capillary electrochromatography, capillary isoelectric focusing, and chiral capillary electrophoresis.

3.038

Jennings, Walter, Eric Mittlefehldt, and Philip Stremple. *Analytical gas chromatography.* 2nd ed. San Diego: Academic Press, 1997. 389 p. 012384357x QD79.C45 543 97-215353

Offers a sound, readable survey of major parts, processes, and considerations in gas chromatography. Treats sample injection, variables, and column selection, as well as special techniques, selected applications, and troubleshooting.

3.039

McMaster, Marvin C. *HPLC, a practical user's guide.*

NY: VCH, 1994. 211 p. 1560816368 QD79.C454 543 93-42139

Written in nontraditional style as a discourse between author and reader, with the unusual inclusion of humor. Brief, simple chromatographic theory; extensive practical material on instrument operation and methodology. Treats purchasing and setting up HPLC instrumentation, development of methodology, troubleshooting and optimization, areas of current application, automation and data acquisition, and future trends.

3.040

McNair, Harold M., and James M. Miller. *Basic gas chromatography.* New York: Wiley-Interscience, 1998. 200 p. (Techniques in analytical chemistry series) 047117260x; 0471172618 pbk QD79.C45 543 97-18151

The introduction is followed by sections discussing stationary phases, capillary columns, detectors, qualitative and quantitative analysis, temperature programming, troubleshooting, and special topics including gas chromatography/mass spectrometry and chiral separations. Appendixes include guidelines for column selection and outline structures, and physical properties of the organic vapor/liquid phases.

3.041

Modern practice of gas chromatography, ed. by Robert L. Grob. 3rd ed. NY: Wiley, 1995. 888 p. (A Wiley-Interscience publication) 0471597007 QD79.C45 543 94-23516

This new edition has changes that reflect the criticism of previous editions. Provides an in-depth, comprehensive treatment of the broad field of gas chromatography. The book is divided into three sections: theory and basics, techniques and instrumentation, and applications. Chapter bibliographies.

3.042

Sadek, Paul Charles. *The HPLC solvent guide.* NY: Wiley, 1996. 346 p. (A Wiley-Interscience publication) 0471118559 QD79.C454 543 96-5235

Covers physical, chemical, chromatographic, and safety issues in liquid chromatography separations. Each of the seven solvent classes is treated in a separate chapter, with tables of physical, chemical, and chromatographic properties; safety and health parameters; availability and chemical structure of solvent. Solvents are grouped by class,

Analytical Chemistry

application, and analyte type. Separations are discussed with sample matrix, list of analytes, chromatographic parameters, and results summaries.

3.043

Snyder, Lloyd R., Joseph J. Kirkland, and Joseph L. Glajch. *Practical HPLC method development*. 2nd ed. NY: Wiley, 1997. 765 p. (A Wiley-Interscience publication) 047100703x QP519.9.H53 543 96-34296

This second edition (1st ed., 1988) of a classic and well-received practical guide on high-performance liquid chromatography (HPLC) contains revised, expanded, and updated material, and recent advances (particularly in areas of computer-assisted method development, analysis of biochemical samples, chiral separations, and preparative HPLC) have been added. Chapters begin with an outline and an introduction and conclude with an extensive list of references and often a bibliography. Clear diagrams, structures, figures, and tables; appendixes with glossary, description of HPLC solvent properties, and guide to preparing buffered mobile phases. Comprehensive cross-referenced index.

Spectroscopy

Reference Works

3.044

Denney, Ronald C. *A dictionary of spectroscopy*. 2nd ed. NY: Wiley, 1982. 205 p. (Wiley-Interscience publication) 0471874787 QC450.3 543 82-4783

Defines about 9,950 words and expressions used in spectroscopy. Terms include the standard vocabulary of this analytical technique, usual laboratory slang, and general scientific terms. Definitions are long paragraphs, some with literature references.

3.045

Encyclopedia of spectroscopy and spectrometry; editor in chief, John C. Lindon; editors, George E. Tranter and John L. Holmes. San Diego: Academic press, 2000. 3 v. 0122266803 (set); 0122266811 (v. 1); 012226682x (v. 2); 0122266838 (v. 3) QC450.3 543 98-87952

A highly specialized but useful set with many cross-referenced items allowing different access points. Entries use short descriptions, references to classics in the field, and cross-references to other, related techniques.

3.046

Handbook of spectroscopy. J.W. Robinson, ed. Cleveland: CRC Press, 1974. May also be listed as CRC handbook ... 3 v. 0878193332 QD95 543 73-77524

Helps to identify materials and compounds using techniques such as infrared, ultraviolet, Raman spectroscopy; electron spin and NMR, emission, atomic, flame, and X-ray spectroscopy; includes mass spectral tables for organics such as ketones, esters, alcohols, and others.

3.047

Herzberg, Gerhard. *Molecular spectra and molecular structure*; with cooperation, in the first edition, of J.W.T. Spinks. 2nd ed. Malabar, FL: R.E. Krieger Pub. Co., 1989-1991. 3 v. Reprinted with corrections. 0894642685 (v. 1) QC454.M6 535.8 88-2933

V. 1: Spectra of diatomic molecules; v. 2: Infrared and Raman of polyatomic molecules; v. 3: Electronic spectra and electronic structure of polyatomic molecules.

3.048

Perkampus, Heinz-Helmut. *Encyclopedia of spectroscopy*, transl. by Heide-Charlotte Grinter and Roger Grinter. Weinheim; New York: VCH, 1995. 669 p. Translation of *Parat Lexikon Spektroskopie*. 3527292810 QD450.3 543 95-7227

Offers information on spectroscopic principles, methods, and applications, ranging from atomic to molecular spectroscopy. Treats methods such as infrared, Raman, NMR, UV-VIS, AAS, ATR, XPS, and others.

Infrared Spectroscopy

3.049

The atlas of near infrared spectra. Philadelphia, PA: Sadtler, 1981. ca. 990 p. 084560063x QC457 543 80-52913

Spectra of 1,000 compounds, listed with name, formula, molecular structure and weight, and other physical properties.

3.050

Bellamy, L.J. *The infrared spectra of complex molecules. v. 2: Advances in infrared group frequencies*. 2nd ed.

London; NY: Chapman & Hall, 1980. 299 p. 0412223503 QC454 80-40094

"The first edition of this book was published with the title *Advances in Infrared Group Frequencies*. This second edition has been retitled to emphasize its relationship to Dr. Bellamy's companion work *The Infrared Spectra of Complex Molecules, v. 1*." Pt. 1: Vibrations of C-C and C-H linkages; alkanes; alkenes; alkynes and allenes; aromatic compounds. Pt. 2: Vibrations involving mainly C-O and O-H linkages; alcohols and phenols; ethers, peroxides, and ozonides; acid halides, carbonates, anhydrides, and metallic carbonyls; aldehydes and ketones; carboxylic acids; esters and lactones. Pt. 3: Vibrations involving mainly C-N and N-H linkages; amides, proteins, and polypeptides; amino acids, their hydrochlorides and salts; amines and imines; unsaturated nitrogen compounds; heterocyclic aromatic compounds; nitro- and nitroso-compounds, nitrates and nitrites; organophosphorus compounds; halogen compounds; organo-silicon compounds; inorganic ions; organosulfur compounds.

3.051

Hershenson, Herbert M. *Infrared absorption spectra.* NY: Academic Press, 1959-1964. QD96 59-7682

V. 1: Index for 1945-1957, 1959, 111 p.; V. 2: Index for 1958-62, 1964, (153 p.) indexes articles having infrared spectra of organic, inorganic, and polymeric compounds, in about 30 journals from 1945-62. Arranged by compound name, referenced to source article. No spectral data are included.

3.052

Nyquist, Richard A. *Interpreting infrared, Raman, and nuclear magnetic resonance spectra.* San Diego: Academic Press, 2001. 2 v. 0125234759 (set) QD96.I5 543 00-108478

V. 1: Variables in data interpretation of infrared and Raman spectra; V. 2: Factors affecting molecular vibrations and chemical shifts of infrared, Raman, and nuclear magnetic resonance spectra.

3.053

Nyquist, Richard A. *The interpretation of vapor-phase infrared spectra.* Philadelphia, PA: Sadtler, 1984. 2 v. 0845600923 (v. 1); 0845601008 (v. 2) QC457 543 83-50549

V. 1: Group frequency data; V. 2: Spectra. Group-frequency data with more than 9,200 IR vapor-phase spectra.

V. 1 includes a general survey of GC/FT-IR and IR spectral search systems. The set contains more than 235 tables and some 500 IR vapor-phase spectra.

3.054

Nyquist, Richard A. *IR and NMR spectral data-structure correlations for the carbonyl group.* Philadelphia, PA: Sadtler Research Laboratories, 1986. 115, [21] p. 0845601350 QD305 546 86-62947

Contains IR and carbon-13 NMR spectra data. Spectra for more than 600 compounds are arranged in 13 tables by chemical class and subclass.

3.055

Pouchert, Charles J. *The Aldrich library of FT-IR spectra.* 2nd ed. Milwaukee: Aldrich Chemical Co., 1997. 3 v., 5,100 p. 094163339x QC457 543 97-73684

Set includes more than 18,000 spectra categorized by chemical functionality and arranged in order of increasing structural complexity. All spectra were run in the Aldrich Quality Control laboratories and confirmed by other spectroscopic and chromatographic analyses. Cross-referenced by chemical name, molecular formula, CASRN, and Aldrich catalog number.

3.056

Pouchert, Charles J. *The Aldrich library of infrared spectra.* 3rd ed. Milwaukee: Aldrich Chemical Co., 1981. 1,850 p. QD96 547.3 81-67533

More than 12,000 spectra classed by functional group and structure.

3.057

The Sadtler handbook of infrared spectra, ed. by William W. Simons. Philadelphia, PA: Sadtler Research Laboratories, 1978. 1,089 p. (The Sadtler handbooks of reference spectra, v. 1) 0845600346 QC453 547 77-95458

Compounds are arranged and indexed by chemical class, compound name, and the Sadtler Spec-Finder system. Information provided includes name; molecular structure, formula, and weight; optical and thermal data; and preparation method.

3.058

Smith, Brian C. *Infrared spectral interpretation: a*

Analytical Chemistry

systematic approach. Boca Raton, FL: CRC Press, 1999. 265 p. 0849324637 QD96.I5 543 98-37190

A short, theoretical introduction is followed by chapters on compounds containing various organic functional groups, inorganic ions, polymers, and electronic and literature aids in interpreting spectra.

3.059

Stuart, Barbara. *Modern infrared spectroscopy*, ed. by David J. Ando. NY: published on behalf of ACOL (University of Greenwich) by Wiley, 1996. 180 p. (Analytical Chemistry by Open Learning) 0471959162; 0471959170 pbk QD96.I5 543 95-32804

A volume in the Analytical Chemistry by Open Learning series. Discusses modern IR techniques, practical applications and techniques, instrumentation, sampling, spectrum interpretation, and quantitative analysis.

Mass Spectroscopy

3.060

Barker, James. *Mass spectrometry*, 2nd ed., ed. by David J. Ando. NY: John Wiley & Sons, 1999. 509 p. (Analytical Chemistry by Open Learning) 0471967645; 0471967629 pbk QD96.M3 543 98-3127

Revised edition of *Mass Spectrometry*, by Reg Davis and Martin Frearson, 1987. Includes fundamental and modern instrumental techniques and self-assessment questions and solutions.

3.061

The Encyclopedia of mass spectrometry, ed. by M.L. Gross and R. Caprioli. Oxford: Elsevier Science, 2003. 10 v., 6,500 p. 0080438504 543.0873

Articles treat basic considerations as well as advanced topics. Contents: v. 1: Theory and ion chemistry; v. 2: Biological applications, Pt. A; v. 3: Biological applications, Pt. B; v. 4: Fundamentals of and applications to organic (and organometallic) compounds; v. 5: Elemental and isotope ratio mass spectrometry; v. 6: Ionization methods; v. 7: Mass analysis and associated instrumentation; v. 8: Hyphenated methods; v. 9: Historical perspective and index; v. 10: Indexes.

3.062

Hoffmann, Edmond de, Jean Charette, and Vincent

Stroobant. *Mass spectrometry: principles and applications*; transl. by Julie Trottier and the authors. Chichester; NY: Wiley; Paris: Masson, 1996. 340 p. 0471966967; 0471966975 pbk QD96.M3 543 96-6746

This book, first published in French in 1994, is the best source of current methods and available instrumentation. About half of the book treats the hardware of mass spectrometry in four chapters: methods of generating ions, mass analyzers, coupling of chromatographs with mass spectrometers, and tandem mass spectrometers. The remainder of the book discusses the interpretation of mass spectra; a chapter is included on the analysis of biomolecules and well-chosen exercises with answers. Illustrations are copious and clear. Twelve appendixes supply a variety of numerical and bibliographic data.

3.063

McLafferty, Fred W., and Douglas B. Stauffer. *The Wiley/NBS registry of mass spectral data*. NY: Wiley, 1989. 7 v. (7,872 p.) (Wiley-Interscience publication) 0471628867 QC454.M3 539 87-31645

A combination of the revisions of the following two books and their data base versions: *Registry of Mass Spectral Data*, by Einar Stenhagen, Sixten Abrahamsson, and Fred W. McLafferty; and *EPA/NIH Mass Spectral Data Base* by S.R. Heller and G.W.A. Milne (and its two supplements).

3.064

McLafferty, Fred W., and Frantisek Turecek. *Interpretation of spectra*. 4th ed. Mill Valley, CA: University Science Books, 1993. 371 p. 0935702253 QC454.M3 543 92-82536

Designed to teach the interpretation of spectra; the material is presented in an orderly, nearly programmed manner. Examples are included.

3.065

McMaster, Marvin C., and Christopher McMaster. *GC/MS: a practical user's guide*. NY: Wiley, 1998. 167 p. 0471248266 QD79.C45 543 97-48529

Uses basic techniques of gas chromatography/mas spectrometry to separate, identify, and quantify individual compounds and substances. Details how to set up and maintain GC/MS systems, perform analyses, and troubleshoot.

3.066

Measuring mass: from positive rays to proteins, ed. by Michael A. Grayson. Philadelphia, PA: Chemical Heritage Press, 2002. 149 p. 0941901319 QD96.M3 543 2001-7646

Originating early in the 20th century, mass spectrometry can determine masses of charged particles in the gas phase. This splendid volume combines an illuminating outline of the history of the technique with examples of its applications, ranging from establishing abundances of isotopes to examining complex biomolecules, including proteins. The first applications of the mass spectrograph involved finding the exact mass and relative abundance of the elements and their isotopes. Recent applications of mass spectrometry extending to planetary space probes, environmental studies, and forensic science are included in the discussion as are techniques such as gas chromatography-mass spectrometry (GC/MS).

3.067

Smith, R. Martin, with Kenneth L. Busch. *Understanding mass spectra: a basic approach.* New York: Wiley, 1999. 290 p. (A Wiley-Interscience publication) 0471297046 QD96.M3 543 98-18136

Smith is a forensic chemist; he contributes to this introduction to interpreting spectra of organic compounds using gas chromatography/mass spectrometry (GC/MS). The book discusses the positive ions formed by electron impact and fragmentation in the gas phase. Instrumentation and other modes of ionization are briefly discussed. There is some discussion of the spectra of illicit drugs such as heroin and their metabolites.

3.068

Sparkman, O. David. *Mass spectrometry desk reference.* Pittsburgh, PA: Global View Pub., 2000. 106 p. 0966081323 QD96.M3 543 00-100995

Terms related to spectroscopy are listed on 64 pages, subdivided into about 15 categories, including data, acquisition methods, instrumentation, and computerized spectra matching. Extensive bibliography of books, spectra collections, and training software.

Nuclear Magnetic Resonance Spectroscopy

3.069

The Aldrich library of 13C and 1H FT-NMR spectra, ed. by Charles J. Pouchert and Jacqlynn Behnke. Milwaukee: Aldrich Chemical Co., 1993. 3 v. 0941633349 QC462.85 547.3 92-73044

Contains 12,000 high-resolution 300MHz 1H and 75MHz 13C FT-NMR spectra, arranged according to functionality. (The electronic version follows.)

3.070

Aldrich/ACD library of FT NMR spectra [computer file]. Milwaukee, WI: Aldrich; Toronto: Advanced Chemistry Development, 1998. 1 CD-ROM QC462.85

Electronic version of *The Aldrich Library of 13C and 1H FT-NMR Spectra*, a printed work described above; it contains C NMR and H NMR spectra of 11,828 organic compounds and information on their physicochemical properties.

3.071

Annual reports on NMR spectroscopy. London; NY: Academic Press, 1970- . ISSN 0066-4103 QC490 538 71-649768

Each new volume contains several papers on recent developments in NMR techniques (1998 is v. 36).

3.072

Bigler, Peter. *NMR spectroscopy: processing strategies.* 2nd updated ed. Weinheim; NY: Wiley-VCH, 2000. 253 p. CD-ROM. (Spectroscopic techniques) 3527288120 QD96.N8 543 00-266560

This book treats all aspects of standard nuclear magnetic resonance (NMR) analysis from data acquisition through structure solution. The accompanying CD-ROM includes special versions of Bruker software programs 1D WIN-NMR, 2D WIN-NMR, and GETFILE plus experimental data for two sample carbohydrates.

3.073

Breitmaier, Eberhard, and Gerhard Bauer. *Carbon-13 NMR spectroscopy: high-resolution methods and applications in organic chemistry and biochemistry.* NY: VCH Publishers, 1987. 515 p. 0895734931 QD96.N8 543 86-28098

Rev. ed. of *13C NMR Spectroscopy*, 2nd ed., 1978. Brief, introductory chapters leading to extended treatment of spectral interpretation, with a problem section with detailed solutions.

3.074

Cowan, Brian P. *Nuclear magnetic resonance and relaxation*. New York: Cambridge University Press, 1997. 434 p. 0521303931 QC762 538 96-46614

Introduces background and general principles of NMR, and includes case studies using NMR imaging. Coverage is highly mathematical.

3.075

Freeman, Ray. *Spin choreography: basic steps in high resolution NMR*. Oxford: Spektrum; Sausalito, CA: University Science Books, 1997. 391 p. 0935702954; 1901217043; 1901217944 RC78.7.N83 543 96-44630

Over the last 20 years, the use of high resolution NMR has become commonplace. The book explains how nuclear spins can be manipulated in order to obtain information about molecules of interest using NMR. Three introductory chapters on energy levels, vector models, and operators provide background. The remaining nine chapters cover topics such as spin echoes, broadband decoupling, 2-D spectroscopy, Nuclear Overhauser Effect (NOE), and coupling constants.

3.076

Friebolin, Horst. *Ein- und Zweidimensionale NMR-Spektroskopie: Basic one-and two-dimensional NMR spectroscopy*, transl. by Jack K. Becconsall. 3rd rev. ed. Weinheim; NY: Wiley-VCH, 1998. 386 p. 3527295135 QP519.9.N83 543 99-184404

This new edition adds pulsed field gradient methods, software for spectra simulation and spin echo experiments. Minimal mathematics; more than 300 illustrations of techniques.

3.077

Macomber, Roger S. *A complete introduction to modern NMR spectroscopy*. New York: Wiley, 1998. 382 p. (A Wiley-Interscience publication) 0471157368 QD96.N8 543 97-17106

A new version of *NMR Spectroscopy: Basic Principles and Applications* (1988). Treats 1H to 13D, 60 to 500 MHz, and one- to two-dimensional spectra, with liquid to solid to living

examples. Also comments on J-J coupling, chemically induced dynamic nuclear polarization, nuclear Overhauser effect, and correlated spectroscopy.

3.078

Pouchert, Charles J. *The Aldrich library of NMR spectra*. 2nd ed. Milwaukee: Aldrich Chemical Co., 1983. 2 v., 2,416 p. QD96 538.362 83-70633

Set includes more than 8,500 60MHz 1H NMR spectra.

3.079

The Sadtler guide to carbon-13 NMR spectra. Philadelphia, PA: Sadtler, 1983. 652 p. 0845600877 QC762 547.3 82-50006

Covers 500 specially selected noise spectra, systematically arranged by compound/carbon type; chemical shift data on an additional 500 compounds are presented in tables. Indexed alphabetically and by heteroatom.

3.080

The Sadtler handbook of proton NMR spectra. William W. Simons, ed. Philadelphia, PA: Sadtler, 1978. 1,254 p. (The Sadtler handbooks of reference spectra, v. 2) 0845600354 QC762 543 78-54281

Compounds are arranged and indexed by chemical class, compound name, and the Sadtler Spec-Finder system. Information provided includes name; molecular structure, formula, and weight; optical and thermal data; and preparation method.

3.081

Sadtler Research Laboratories. *The Sadtler handbook of ultraviolet spectra*. William W. Simons, ed. Philadelphia, PA: Sadtler Research Laboratories, 1979. 1,016 p. (The Sadtler handbooks of reference spectra, v. 3) 0845600338 QC459 547.3 79-65539

Compounds are arranged and indexed by chemical class, compound name, and the Sadtler Spec-Finder system. Abridged edition of the Sadtler standard ultraviolet spectra collection. Information provided includes name; molecular structure, formula, and weight; optical and thermal data; and preparation method.

3.082

Simons, William W., and M. Zanger. *The Sadtler guide*

to NMR spectra. Philadelphia, PA: Sadtler Research Laboratories, 1972. 542 p. 084560001x QC490 547 72-75379

Compounds are selected from the Sadtler Standard NMR Collection. Divided into 12 classifications, providing chemical shift, coupling constant, and other significant features. Some 480 compounds, including 130 compounds that have a single spectral peak, are listed.

3.083

Ultraviolet and visible absorption spectra, ed. by Herbert M. Hershenson. NY: Academic Press, 1956-66. 3 v. QD95 56-8684

V. 1: Index for 1930-1954; v. 2: Index for 1955-1959. V. 3 indexes optical rotatory dispersion and circular dichroism spectra. Separate volume for index to 1960/1963 published in 1966 covers organic, inorganic, and polymeric compounds from some 70 journals. Arranged by substance name with reference to source article; no spectral data are included.

Ultraviolet Spectroscopy

3.084

UV spectroscopy: techniques, instrumentation, data handling: UV Spectrometry Group; ed. by B.J. Clark, T. Frost, and M.A. Russell. London; NY: Chapman & Hall, 1993. 146 p. (Techniques in visible and ultraviolet spectrometry; v. 4) 041240530x QD96.U4 543 92-47358

Contents: Fundamental principles; Standards; Instrumentation for UV spectroscopy and fluorescence: multicomponent analysis, Spectral libraries, Colour; Diode array detection in HPLC: Validation, Supplementary experiments; Appendix.

Other Spectroscopic Techniques

3.085

Alkemade, C. Theodorus J., and R. Herrmann. *Fundamentals of analytical flame spectroscopy*; translated from German by R. Auerbach and Paul T. Gilbert, Jr. NY: Wiley, 1979. 442 p. (A Halsted Press book) 0470267100 QD96.F5 543 79-4376

Good source of authoritative information, with extensive bibliographies; writing and illustrations are clear. Glossary; atlas of flame spectral lines.

3.086

Barr, Tery Lynn. *Modern ESCA: the principles and practice of X-ray photoelectron spectroscopy.* Boca Raton, FL: CRC Press, 1994. 358 p. 0849386535 QD96.E44 543 93-26854

Describes use of core level and valence band binding energies, their shifts, and line widths; background, present status, and possible future uses of branches of electron spectroscopy for chemical analysis (ESCA). New analyses are described for various oxides, inert hydrocarbon polymers, carbon-filled metal ceramics, supported metals catalysts, unique structures, and other systems. Offers a general description of the use of ESCA to analyze high-T_c superconducting thin film and leaves of trees during senescence.

3.087

Hollas, J. Michael. *Basic atomic and molecular spectroscopy.* NY: Wiley-Interscience; Cambridge: Royal Society of Chemistry, 2002. 184 p. (Basic concepts in chemistry) 047128162x QC451

Explains theory, techniques, and relationships between different spectroscopic techniques, such as rotational, vibrational, and electronic spectroscopy. Encompasses both high resolution (structural) and low resolution (analytical) techniques.

3.088

Hollas, J. Michael. *High resolution spectroscopy,* 2nd ed. Chichester; NY: J. Wiley, 1998. 743 p. 0471974218 QC454 543 97-44183

Thorough update from first edition of 15 years ago; increased material on laser spectroscopy; coverage of rotational, vibrational, electronic, and photoelectronic methods; more tables, figures, and references.

3.089

Jenkins, Ron. *X-ray fluorescence spectrometry.* 2nd ed. NY: Wiley, 1999. 207 p. (A Wiley-Interscience publication) 0471299421 QD96.X2 543 98-39008

Discusses the history of X-ray fluorescence spectroscopy, design of X-ray spectrometers, X-ray spectra, and state-of-the-art applications. Techniques and procedures for using this application in quantitative and qualitative analysis are presented.

3.090

Modern techniques in applied molecular spectroscopy, ed. by Francis M. Mirabella. NY: Wiley, 1998. 410 p. (A Wiley-Interscience publication) 0471123595 QD96.M65 543 97-13437

Contents: Transmission infrared spectroscopy; Specular reflection spectroscopy; Attenuated total spectroscopy; Photoacoustic spectroscopy; Infrared microspectroscopy; Raman microspectroscopy; Emission spectroscopy; Fiber optics in molecular spectroscopy.

3.091

Particle-induced X-ray emission spectrometry (PIXE), ed. by Sven A.E. Johansson, John L. Campbell, and Klas G. Malmqvist. NY: Wiley, 1995. 451 p. (Chemical analysis, v. 133) (A Wiley-Interscience publication) 0471589446 QD96.X2 545 94-44471

Successfully incorporates basic principles in a very readable format. A more comprehensive discussion of fundamentals follows; theory concludes with a discussion of the high energy microprobe modification and applications of PIXE, including biological and medical applications, atmospheric aerosols, PIXE in the earth sciences, and a unique chapter concerned with archaeology and art and how PIXE can be used to characterize artifacts and art works. The book concludes with a discussion on how PIXE stacks up against other elemental analysis methods.

3.092

Van Loon, Jon C. *Analytical atomic absorption spectroscopy: selected methods.* NY: Academic Press, 1980. 337 p. 0127140506 QD96 543 79-25448

Includes bibliographic references and index.

Polarography

3.093

Thomas, F.G., and G. Henze. *Introduction to voltammetric analysis: theory and practice.* Collingwood, Victoria, Australia: CSIRO Pub., 2001. 252 p. 0643065938 QD116.V4 543.0871

Translation of *Polarographie und Voltammetrie: Grundlagen und Analytische Praxis.* This book is small and far from comprehensive, but there is sufficient introductory background material on electrochemical principles that students could learn enough on their own to get started on an electrochemical project. The book adds specific experimental directions for a variety of typical analyses, and will be useful for self-guided laboratory projects.

Treatises

3.094

Analytical chemistry: the approved text to the FECS curriculum analytical chemistry, ed. by R. Kellner [et al]. Weinheim; NY: Wiley-VCH, 1998. 916 p. 3527288813; 3527286101 pbk QD101.2

Offers to newcomers as well as students a modern, structured overview of analytical chemistry, worldwide.

3.095

Comprehensive analytical chemistry, ed. by Cecil L. Wilson and David W. Wilson. Amsterdam; NY: Elsevier, 1959-<1992 >. v. <1A-C, 2A, 2C-D, 6-7, 9-11, 12A-E, 13-22, 28-29 > QD75 543 58-10158

V. 1A,B,C: Classical analysis; v. 2A,B,C,D: Electrical methods, physical separation methods; v. 3: Optical methods. V. 4: Industrial applications. V. 5: Miscellaneous; general index; v. 19: Analytical visible and ultraviolet spectrometry; v. 20: Photometric methods in inorganic trace analysis; v. 28: Kinetic methods in chemical analysis, application of computers in analytical chemistry; v. 31: Analytical and biomedical applications of ion-selective field-effect transistors. Later accounts have noted that there are other volumes: v. 6-7, 9-11, 12A-E, 13-22. G. Svehla edited v. 9-10, 12A-12C, 12E, 14, 21-22, under the title: *Wilson and Wilson's Comprehensive Analytical Chemistry.*

3.096

Treatise on analytical chemistry, ed. by Izaak M. Kolthoff and Philip J. Elving. NY: Wiley-Interscience, 1959- . 3 parts in v. 0471499668 (pt.1, v. 10) QD75 543 59-12439

Three parts: Pt. I. (6 v) Theory and practice of analytical chemistry; Pt. II. (13 v) Analytical chemistry of the elements; Pt. III. (3 v) Analysis of industrial products (v. 1 ed. by I.M. Kolthoff, P.J. Elving, and F.H. Stross). The 2nd ed. was started in 1978, with various volumes and parts being updated at irregular intervals (older edition had 17 volumes). Comprehensive treatment of the theoretical fundamentals of analytical chemistry and implementation, as well as the practice of inorganic and organic analysis.

3.097
Treatise on analytical chemistry. 2nd ed. Izaak M. Kolthoff and Philip J. Elving, eds. NY: Wiley, 1978-c1989 >- . Pt. 1, v. 1-2, 4-5, 7-8, 11-12, 047103438x (pt.1, v. 1) QD75.2 543 78-1707

Pt. 1: Theory and practice. Pt. 1, v. 8- ed. by Philip J. Elving; assoc. ed., Edward J. Meehan. Pt. 1, v. 11 ed. by James D. Winefordner; assoc. ed. Maurice M. Bursey. Includes bibliographies and indexes.

3.098
Valcárcel Cases, Miguel. *Principles of analytical chemistry: a textbook*. Berlin; New York: Springer, 2000. 371 p. 354064007x QD75.2 543 00-33829

Overviews the field of analytical chemistry; includes information, chemical references; essential features of analytical chemistry; evolution of this field; conceptual and technical hierarchies; and classifications.

3.099
Wang, Joseph. *Analytical electrochemistry*. 2nd ed. New York: John Wiley & Sons, 2000. 209 p. 0471282723 QD115 543 99-89637

Includes many new topics, such as self-assembled monolayers, DNA biosensors, single molecule detection, and micromachined analyzers. Full range of electrochemical techniques and devices is discussed.

SAMPLING AND STATISTICAL METHODOLOGY

3.100
Beebe, Kenneth R., Randy J. Pell, and Mary Beth Seasholtz. *Chemometrics: a practical guide*. NY: Wiley, 1998. 348 p. 0471124516 QD75.4.S8 543 97-17970

Chemometrics uses statistical tools to design, analyze, and interpret experimental data. This book covers preprocessing, multivariate calibration, and pattern recognition.

3.101
Crawford, Karen, and Alan Heaton. *Problem solving in analytical chemistry*, compiled and developed by Karen Crawford and Alan Heaton; ed. by Denise Rafferty and

Sara Sleigh. London: Education Division, the Royal Society of Chemistry, 1998. 145 p. 1870343468 pbk 543

A variety of problems are featured, complete with solutions and guides for tutors. Subjects range from gravimetric analysis to interpretation of spectroscopic data.

3.102
Meloan, Clifton E. *Chemical separations: principles, techniques, and experiments*. New York: Wiley, 1999. 752 p. (A Wiley-Interscience publication) 0471351970 QD63.S4 543 99-36208

A comprehensive and practical work on separation, covering numerous techniques used to prepare materials for chemical analysis.

3.103
Otto, Matthias. *Chemometrics: statistics and computer applications in analytical chemistry*. Weinheim; New York: Wiley-VCH, 1999. 314 p. 352729628x QD75.4.S8 543 99-186538

Treats descriptive and inferential statistics; popular methods of chemometrics, including experimental design, signal processing, pattern recognition, data banks, knowledge processing, and quality assurance and good lab practice.

HISTORY AND BIBLIOGRAPHY

3.104
A century of separation science, ed. by Haleem J. Issaq. NY: Marcel Dekker, 2002. 755 p. 0824705769 QD79.C4 660 2001-59229

Separately authored chapters by experts in chromatography, electrophoresis, field-flow fractionation, countercurrent chromatography, and supercritical fluid chromatography.

3.105
A history of analytical chemistry, ed. by Herbert A. Laitinen and Galen W. Ewing. Washington, DC: Div. of Analytical Chemistry of the American Chemical Society, 1977. 358 p. QD72 543 77-23923

Historical overview of the field.

PROBLEM MANUALS AND SOFTWARE

3.106
Gordus, Adon A. *Schaum's outline of theory and problems of analytical chemistry.* NY: McGraw-Hill, 1985. 242 p. 0070237956 QD101.2 545 84-27823

Contains 590 fully solved problems and others with solutions supplied for student testing. Explanations are provided for basic concepts and mathematical principles of analytical chemistry.

3.107
Kenkel, John. *Analytical chemistry for technicians.* 3rd ed. Boca Raton, FL: Lewis Publishers, 2003. 554 p. 1 CD-ROM 1566705193 QD75.22 543 2002-29654

Contents: Introduction; Sampling and sample preparation; Gravimetric analysis; Titrimetric analysis; Instrumental analysis; Spectrochemical analysis; UV and IR molecular spectrometry; Atomic spectroscopy; Other spectroscopic methods; Analytical separations; Gas chromatography; High performance liquid chromatography; Electroanalytical methods; Physical testing methods; Bioanalysis; and Good laboratory practices.

Chapter 4

Physical Chemistry

In this part of our adventure through the chemical literature, the portion of this field occupied by the literature of physical chemistry is treated. This area is where one finds information on physical constants, physical property data, and related material that is specific to the physical behavior of chemical compounds. However, any treatment of this type that concentrates specifically on inorganic or organic chemistry, etc., will be placed in the book with materials for that part of chemistry. For instance, resources treating electrochemistry, without mention of type of chemistry or technique, will be treated in this chapter. As in other sections of chemistry, there are treatises, handbooks, dictionaries, tables of all types of data, and information to aid the searcher; here we will discuss this literature. However, keep in mind that anything discussed in previous chapters of this book will also be useful, even though not mentioned specifically here (such as general encyclopedias or handbooks in chemistry or chemical engineering).

LITERATURE RETRIEVAL

4.001
Arny, Linda Ray. *The search for data in the physical and chemical sciences*. NY: Special Libraries Association, 1984. 150 p. 0871113802 pbk QC5.3 500.2 83-20376

Contents: The creation, compilation, and retrieval of data; Data compilations of the National Bureau of Standards; appendixes. Discusses nature of data, locating data in the scientific literature, critical evaluation of data, data centers, history of data compilations, selection sources for handbook and data compilations, data journals and related publications, the National Bureau of Standards, National Standard Reference Data System, and others.

4.002
Journal of physical and chemical reference data. Washington; NY: American Chemical Society, v. 1- , 1972- . Quarterly; bimonthly, 1992- . ISSN 0047-2689 Q199 530 72-622555

Forms part of National Standard Reference Data System of the National Bureau of Standards. Supplements accompany some volumes. Published on behalf of NBS.

4.003
Physical chemistry source book; ed. in chief, Sybil P. Parker. NY: McGraw-Hill, 1988. 406 p. (The McGraw-Hill Science Reference Series) 007045504x QD451 541.3 87-36629

Includes bibliographies. A single source of understandable explanations for physical chemical phenomena. Sections treat thermodynamics, kinetics, surface chemistry, transport properties, and electrochemistry.

ENCYCLOPEDIAS

4.004
Encyclopedia of computational chemistry; ed. in chief, Paul von Ragué Schleyer. Chichester; NY: J. Wiley, 1998. 5 v., 3,429 p. 047196588x QD39.3.E46 542 98-37164

This set provides information concerning modeling by computation of all aspects of chemistry. Because computational chemistry is a helpful and desirable tool, this reference offers comprehensive, up-to-date coverage of this rapidly developing field of chemistry. There are three types of entries: regular articles, definition entries that explain the common terminology of the field, and descriptions of available software packages. The application of computer methods to analyze chemical information is interpreted broadly, and as a result topics cover a wide spectrum, including molecular structure computations, chemometrics, chemical engineering applications, infrared data correlations, laboratory information management systems, simulations of biological and organic molecules, free energy calculations, molecular dynamics, NMR correlations, and teaching of computational chemistry. Complete table of contents for all volumes; entries have appropriate background and references; figures and photographs are clear and properly labeled. Extensive list of references; extensive list of selected software.

4.005

The International encyclopedia of physical chemistry and chemical physics, ed. by E.A. Guggenheim, J.E. Meyer, and F.C. Tompkins. NY: Pergamon; Macmillan, 1965- . 22 v. QD453 541.082 65-7513

DICTIONARY

4.006

Comprehensive dictionary of physical chemistry, ed. by L. Ulický and T.J. Kemp. NY: E. Horwood; PTR Prentice Hall, 1992. 472 p. (Ellis Horwood PTR Prentice Hall physical chemistry series) 0131517473 QD5 540 91-24322

Covers classical areas such as thermodynamics, electrochemistry, chemical statistics, kinetics, optical spectroscopy, molecular properties, and surface and colloid chemistry, as well as nuclear and solid state physics, group theory, magnetic resonance spectroscopy, photophysics, and photochemistry. Includes theoretical and quantum chemistry, symmetry, and crystallography.

GENERAL HANDBOOKS

4.007

Condon, Edward U., and Hugh Odishaw. *Handbook of physics.* 2nd ed. NY: McGraw-Hill, 1967. 1,626 p. QC21 530 66-20002

Standard handbook with chapters by specialists; encyclopedic style. Sections discuss mathematics, mechanics of particles and rigid bodies, mechanics of deformable bodies, electricity and magnetism, heat and thermodynamics, optics, atomic physics, solid state, nuclear physics. Bibliography.

4.008

Lide, David R., and Henry V. Kehiaian. *CRC handbook of thermophysical and thermochemical data.* Boca Raton, FL: CRC Press, 1994. 518 p.; 1 disk. 0849301971 QC173.397 541.3 93-36909

Comprehensive tabulation of thermodynamic and transport properties of pure substances and mixtures. Transport properties listed include viscosity, thermal conductivity, and diffusion coefficients.

DATABASES AND LITERATURE REVIEWS

4.009

DIPPR (AIChE Design Institute for Physical Property Data). *Data compilation of pure compound properties database.* American Institute of Chemical Engineers; Office of Standard Reference Data, National Bureau of Standards, Gaithersburg, MD. v. 1- , 1991- . 2-19467

Online through STN or for stand-alone systems. As of 1992, the database contains information on 39 properties for 1,212 chemicals of high industrial priority; includes thermodynamic, physical, and transport properties of each chemical, and for each, values are given for 26 single-valued property constants and for 13 properties as functions of temperature calculated from correlation coefficients. Included also are estimates of the accuracy of each property value and references to the sources of measured or predicted data used in the selection of recommended values.

4.010

Electronic Materials Information Service (EMIS) database. Stevenage, Herts, UK: INSPEC, 1980- . ISSN 0950-1398 89-11041

Available through various online vendors. Provides physical properties data on more than 250 semiconductor materials; more than 800 reviews by 250 experts included.

TREATISES, MONOGRAPHS, AND ADVANCED TEXTBOOKS

4.011

Advances in chemical physics, ed. by Ilya Prigogine et al. NY: Wiley; v. 1, 1958- . Irregular. ISSN 0065-2385 QD453 541 58-9935

V. 81, 1992; v, 1 (1958)-45 (1981) as v. 59; v. 109 was published in 1999. Reports recent advances in every area of chemical physics. Significant, up-to-date chapters by internationally recognized researchers offer comprehensive analyses of areas of interest.

4.012

Annual review of physical chemistry. Palo Alto, CA: Annual Reviews, v. 1- , 1950- . ISSN 0066-426x QD1 541.058 51-1658

Lengthy reviews on various important topics in physical

chemistry, often reviewing a number of years' work on a specific topic. Lengthy bibliographies for each article. Book-length, hardbound volume.

4.013

Bernasek, Steven L. *Heterogeneous reaction dynamics*. NY: VCH, 1995. 158 p. 0895737426 QD461 541.3 95-14954

Briefly summarizes experimental techniques and theory used in studying the interactions at the gas-solid interface, including both nonreactive ("physical" processes such as energy transfer, nonreactive adsorption, and diffusion) and the reactive events characteristic of surface catalysis. Selected illustrative examples from the literature; clear writing; recommended as an introduction to the subject. Chapter headings: "Surface Characterization Methods"; "Inelastic Scattering and Energy Transfer"; "Adsorption, Epitaxial Growth, and Adsorbate Interactions"; "Surface Diffusion"; "Dynamics of Dissociative Adsorption"; "Atom Recombination Dynamics"; "Catalytic Oxidation"; and "Small Molecule Decomposition Processes."

4.014

Berry, R. Stephen, Stuart A. Rice, and John Ross. *Physical chemistry*. 2nd ed. NY: Oxford University Press, 2000. 1,064 p. (Topics in physical chemistry) 0195105893 QD453.2 541.3 00-24923

Discusses molecular structure, equilibrium properties of systems, and kinetics of transformations of systems, from a standpoint of, respectively, quantum mechanics, thermodynamics and equilibrium statistical mechanics, and chemical kinetics and kinetic theory.

4.015

Bromberg, J. Philip. *Physical chemistry*. 2nd ed. Boston: Allyn and Bacon, 1984. 993 p. 0205080197 QD453.2 541.3 83-21531

A useful work, oriented toward environmental and energy conservation considerations.

4.016

Drago, Russell S. *Physical methods in chemistry*. Philadelphia: Saunders, 1977. 660 p. 0721631843 QD453.2 543 76-8572

(1965 ed.: *Physical Methods in Inorganic Chemistry*.)

4.017

Gil, Victor M.S. *Orbitals in chemistry: a modern guide for students*. Cambridge; NY: Cambridge University Press, 2000. 314 p. 0521661676; 052166649X pbk QD461 541.2 99-461968

An introductory, primarily qualitative, discussion of the role of atomic and molecular orbitals in chemistry. Not as demanding mathematically as traditional works on quantum chemistry. Introduces fundamentals of quantum mechanics, atomic and molecular orbitals for one-electron systems, molecular orbitals and bonding in diatomic and polyatomic molecules, and direct relations between orbitals and chemical reactivity and orbitals and spectroscopy.

4.018

Glasstone, Samuel S. *Textbook of physical chemistry*. 2nd ed., 1940. Princeton, NJ: Van Nostrand, 1965. 1,320 p. QD453 541 40-32831

Contents: Atomic structure and atomic spectra; Radioactivity and isotopes; First and second laws of thermodynamics; The gaseous state; The solid state; Changes of state; The liquid state; Physical properties and molecular structure; Dilute solutions; Phase equilibria; Chemical equilibrium and free energy; Electrochemistry; Chemical kinetics; Surface phenomena.

4.019

Hargittai, István, and Magdolna Hargittai. *Symmetry through the eyes of a chemist*. 2nd ed. NY: Plenum Press, 1995. 469 p. 03064488513; 0406448521 pbk QD461 541.2 95-30533

(1st ed., 1987.) Includes examples of symmetry of human invention and natural origin found in travels throughout the world, and in literature both ancient and modern, coupled with a comprehensive treatment of the uses of symmetry in modern chemistry. The book's strengths include the many photographs of objects with distinctive symmetry, the nonmathematical presentation of point group symmetry and its applications to inorganic chemistry and organic reaction mechanisms, and extensive chapter bibliographies.

4.020

Kettle, S.F.A. *Symmetry and structure*. Chichester, UK; NY: Wiley, 1985. 330 p. 0471905011; 0471907057 pbk QD471 541.2 84-17365

Elementary work on group theory as applied to chemical problems, with emphasis on bonding and electronic structure. Appendixes contain most of the formal theory. New examples are provided of group theory applied to bonding and electronic structure. Numerous problems; many good figures. Reprinted with corrections in 1986.

4.021

Moelwyn-Hughes, E.A. *Physical chemistry.* 2nd rev. ed. NY: Pergamon, 1965. 1,334 p. QD453 541 66-4343

(1st ed., 1947.) General principles of physical chemistry; applications to certain simple systems. Edition of 1957: Kinetic molecular theory; quantum theory; chemical elements; chemical thermodynamics; intermolecular energy; partition functions; molecules and light; crystalline, metallic, liquid, ionic interfacial states; chemical equilibria; reaction kinetics.

4.022

Moore, Walter John. *Physical chemistry.* 4th ed. Englewood Cliffs, NJ: Prentice-Hall, 1972. 977 p. 0136659683 QD543.2 541 70-156983

Twenty chapters include physicochemical systems, energetics, entropy and free energy, kinetic theory, statistical mechanics, changes of state, solutions, chemical affinity, chemical reaction rates, electrochemistry/ionics, interfaces, electrochemistry/electrodics, particles and waves, quantum mechanics and atomic structure, the chemical bond, symmetry and group theory, spectroscopy and photochemistry, the solid state, intermolecular forces and the liquid state, and macromolecules.

4.023

Nonhebel, D.C., J.M. Tedder, and J.C. Walton. *Radicals.* Cambridge, UK; NY: Cambridge University Press, 1979. 200 p. 05221220041; 0521293324 pbk QD471 541 78-54721

Shorter, more readable version of the authors' *Free-radical Chemistry* (1974). Includes material on radicals in biological systems, electron spin resonance spectroscopy, and chemically induced dynamic nuclear polarization.

4.024

Physical chemistry, an advanced treatise, ed. by Henry Eyring, Douglas Henderson, and Wilhelm Jost. NY: Academic Press, 1967-75. 11 v. ISSN 1045-6082 QD453 541.3 89-7775

V. 1: Thermodynamics; v. 2: Statistical mechanics; v. 3: Electronic structure of atoms and molecules; v. 4: Molecular properties; v. 5: Valency; v. 6: General introduction and gas reactions; v. 7: Reactions in condensed phases; v. 8: Liquid state; v. 9A, 9B: Electrochemistry; v. 10: Solid state; v. 11: Mathematical methods. V. 11B publ. in 1975. (Individual volume editors.)

4.025

Physical methods of chemistry, ed. by Bryant C. Rossiter and John F. Hamilton. 2nd ed. NY: Wiley, 1986-1993. 10 v. in 12. 0471080349 (v. 1); 0471080276 (v. 2); 0471850411 (v. 3A); 0471850519 (v. 3B); 0471080268 (v. 4); 047152509x (v. 5); 0471570877 (v. 6); 0471534382 (v. 7); 0471544078 (v. 8); 047154406x (v. 9A); 0471544051 (v. 9B); 0471570869 (v. 10) QD61 542 85-6386

V. 1: Components of scientific instruments and applications of computers to chemical research; v. 2: Electrochemical methods; v. 3: Determination of chemical composition and molecular structure (2 v.); v. 4: Microscopy; v. 5: Determination of structural features of crystalline and amorphous solids; v. 6: Determination of thermodynamic properties; v. 7: Determination of elastic and mechanical properties; v. 8: Determination of electronic and optical properties; v. 9A-B: Investigations of surfaces and interfaces (2 v.); v. 10: Supplement and cumulative index.

4.026

Pilling, Michael J., and Paul W. Seakins. *Reaction kinetics.* NY: Oxford University Press, 1996. 305 p. (Oxford science publications) 0198555288; 019855527x pbk QD502 541.3 94-46221

(Reprinted with corrections, 1997) Ranges through experimental methods in gas phase, at metal surfaces, and in solutions. Theoretical models are developed from many perspectives: transition-state theory, collision theory, the Lindemann unimolecular reaction theory, and RKK-M modifications. Especially interesting and helpful is the compilation and presentation of modern methods of measurement of kinetics—from laser-induced reactions, reactions in the atmosphere, and enzyme kinetics to the field of reaction dynamics. The book replaces other kinetics works (even that of John W. Moore and Ralph G. Pearson, *Kinetics and Mechanism*, 3rd ed., 1982) because it truly covers modern kinetics logically, carefully, mathematically, and succinctly.

4.027

Rappé, Anthony K., and Carla J. Casewit. *Molecular*

mechanics across chemistry. Sausalito, CA: University Science Books, 1997. 444 p. 0935702776 QD461 541.2 96-13315

The goal of molecular mechanics is the understanding of properties of molecules and materials by modeling them as physical (mechanical) systems, taking as given whatever can be determined about the properties of the component atoms and the force fields that bond them together. Modern computing techniques have opened many very complex systems (macromolecules, complex crystals, liquids) to semiempirical molecular mechanics analysis, so that the structures and chemical functions of substances as complex as DNA, proteins, dissolved polymers, and zeolites may now be predicted, understood, and compared to experimental observations, with reference to quantitative models.

4.028

Seddon, John M., and Julian D. Gale. *Thermodynamics and statistical mechanics*. Cambridge: Royal Society of Chemistry, 2001. 161 p. (Tutorial chemistry texts, 10) 0854046321 QD504 536.7

Thermodynamic quantities and relationships are introduced and developed, enabling students of apply thermodynamic analysis to chemical problems. Each stage in the development is well illustrated with examples, helping students to understand energy, its different forms and transformations, and the key role of entropy as applied to chemical systems.

4.029

Silbey, Robert J., and Robert A. Alberty. *Physical chemistry*. 3rd ed. NY: Wiley, 2001. 969 p. 0471383112 QD453.2 541.3 00-20734

(2nd ed., 1997.) Intended for undergraduates as a textbook for physical chemistry courses taken by chemistry majors, chemical engineers, and biochemists as upper-level undergraduates. Features very clear explanations of concepts and methods, together with problem solving and Mathematica-based problems.

4.030

Tabor, David. *Gases, liquids and solids: and other states of matter*. 3rd ed. Cambridge; NY: Cambridge University Press, 1991. 418 p. 0521404886; 0521406676 pbk QC173 530.4 91-9214

(2nd ed., 1979; 1st ed., 1969) A well-written text in physical chemistry, suitable for undergraduates.

4.031

Techniques of chemistry, ed. by Arnold Weissberger. NY: Wiley; 1971-<c1984 > v. <18 > in <29 >. Irregular. 0471932663 (v. 8, pt. 1) ISSN 0082-2531 QD61 542 85-649508

Supersedes: *Technique of Organic Chemistry* and its companion, *Technique of Inorganic Chemistry*, ed. by H. B. Jonassen and A. Weissberger. This series of volumes is by experts and covers the various techniques of synthetic work.

4.032

Verkade, John G. *A pictorial approach to molecular bonding and vibrations*. 2nd ed. NY: Springer, 1997. 367 p. 0387948112 QD461 541.2 96-19135

(Rev. ed. of *A Pictorial Approach to Molecular Bonding*, 1986.) Treats the formulation and visualization of molecular orbitals and molecular vibrations created by applying the "generator orbital" method to a minimum basis set of atomic orbitals. Equivalent orbital transformations and relations of orbital hybridization schemes to canonical molecular orbitals are discussed. Chapters treat the orbital picture for bound electrons in atoms, diatomic molecules, and categories of polyatomic systems.

4.033

Warren, Warren S. *The physical basis of chemistry*, 2nd. ed. San Diego: Academic Press, 2000. 211 p. (Complementary science series) 0127358552 QD475 541 99-68993

Filled with brilliant scientific insights and interesting and clever anecdotes, this book will extend a reader's understanding of physical chemical concepts, and there are examples to introduce differentiation, chain rule, integrals, and how they are used to solve actual problems encountered in physical chemistry. Newton's Laws of Motion, forces, kinetics, properties of waves, statistical mechanics, quantum mechanics, and the interaction of radiation and matter are discussed.

4.034

Weissberger, Arnold. *Physical methods of chemistry*, incorporating 4th completely rev. and augmented ed. of *Technique of organic chemistry*, ed. by Arnold Weissberger and Bryant C. Rossiter. NY: Wiley-Interscience, 1971-1977. 6 v. in 11. 0471927244 (v. 1) QD61 542 75-29544

Pt. 1A: Components of scientific instruments. Pt. 1B:

Automatic recording and control, computers in chemical research. Pt.2: Electrochemical methods (2 v.). Pt.3: Optical, spectroscopic, and radioactivity methods: A. Interferometry, light scattering, microscopy, microwave, and magnetic resonance spectroscopy. B. Spectroscopy and spectrometry in the infrared, visible, and ultraviolet C. Polarimetry D. X-ray, nuclear, molecular beam, and radioactivity methods (4 v.). Pt.4: Determination of mass, transport, and electrical-magnetic properties. Pt.5: Determination of thermodynamic and surface properties. Pt.6: Supplement and cumulative index.

Thermodynamics

4.035

Anderson, Gregor Munro. *Thermodynamics of natural systems*. NY: Wiley, 1996. 382 p. 0471109436 QE515.5.T46 541.3 95-23040

Aimed at areas such as forestry, soil science, or other disciplines where the thermodynamics of natural systems is important. Thermodynamics is applied throughout: fundamental laws of thermodynamics are introduced; familiar examples are cited; and helpful analogies to mechanical systems are used. Thermodynamic principles are applied in later chapters, which treat topics from calorimetry to statistical thermodynamics. Appendix with review of principles of calculus.

4.036

Dugdale, J.S. *Entropy and its physical meaning*. London: Bristol, PA: Taylor & Francis, 1996. 198 p. 0748405682; 0748405690 pbk QC318.E57 536 96-218083

An easy-to-read introduction to entropy and its connection to the driving forces present in physical and chemical changes. Explains the development of scientists' understanding of entropy and the statistical interpretation of the concept using many interesting references to experimental work. Although much of the coverage is similar to that seen in standard physical chemistry textbooks, this book provides more complete and coherent explanations of what are often difficult concepts for undergraduate students.

4.037

Hinchliffe, Alan. *Chemical modeling: from atoms to liquids*. Chichester; NY: Wiley, 1999. 395 p. 0471999032; 0471999040 pbk QD455.3.C64 541 99-32077

Presents a comprehensive overview of the background and applications of modern chemical modeling approaches. Describes classical, quantum, and statistical mechanics at a basic level, leading to an up-to-date treatment of solid state theory, molecular mechanics, molecular dynamics, quantum methods, and simple approaches to modeling polymers and liquids. Commercially available desktop computer software was used to generate many illustrative examples and figures.

4.038

Honig, Jurgen. *Thermodynamics*. 2nd ed. San Diego: Academic Press, 1999. 608 p. 0123550459 QD504 541.3 98-86761

Provides a concise overview; first principles, postulations, and self-contained descriptions of physical and chemical processes.

4.039

Jeffrey, George A. *An introduction to hydrogen bonding*. NY: Oxford University Press, 1997. 303 p. (Topics in physical chemistry) 0195095480; 0195095499 pbk QD461 541.2 96-26792

Provides much useful information, such as characteristic geometries for particular hydrogen bonding patterns, and contains conversion factors between commonly used units for energy and distance. The book offers a brief history of the subject, which provides insight into the problems that are faced when defining and explaining hydrogen bonding. Chapters discuss the nature and properties of the hydrogen bond as determined from experimental and theoretical studies; patterns of hydrogen bonds in crystal structures and chapters on the geometry of various hydrogen bonding schemes as determined by crystal structure data; hydrogen bonding as it appears in particular classes of compounds such as clathrates, inclusion compounds, macromolecules, hydrates, and ices.

4.040

Kondepudi, Dilip K., and Ilya Prigogine. *Modern thermodynamics: from heat engines to dissipative structures*. Chichester; NY: John Wiley & Sons, 1998. 486 p. 0471973939; 0471973947 pbk QD311 536 97-48745

Presents thermodynamics in a contemporary manner, with applications, exercises, and real-life examples.

4.041

Kudryavtsev, Andrei B., Reginald F. Jameson, and

Wolfgang Linert. *The law of mass action*. Berlin; NY: Springer, 2001. 328 p. 3540410783 QD503 541.3 2001-17048

Discusses Maxwell-Boltzmann statistics; ensembles, partition functions, and thermodynamic functions; the law of mass action for ideal systems; reactions in imperfect condensed systems; free volume; molecular interactions; imperfect gases; reactions in imperfect condensed systems; lattice energy; and chemical correlations.

4.042

McQuarrie, Donald A., and John D. Simon. *Molecular thermodynamics*. Sausalito, CA: University Science Books, 1999. 656 p. 189138905x QD504 541.3 98-48543

Based on part of the recently published textbook *Physical Chemistry—A Molecular Approach*. Some new material and three entirely new chapters are added. Several "Math Chapters" help readers review important topics that are used extensively in subsequent chapters. A particularly useful organizational feature is that each chapter is broken down into several sections, each headed by a "take home" topic sentence. A molecular view of thermodynamics is stressed, beginning with a chapter summarizing some important results from quantum mechanics that lead naturally into important thermodynamic principles, such as the Boltzmann factor, partition functions, and the three laws of thermodynamics. Chapters consider phase equilibria, solutions, chemical equilibrium, and electrochemical cells; and nonequilibrium (irreversible) thermodynamics. Large number of worked examples and end-of-chapter problems.

4.043

McQuarrie, Donald A., and John D. Simon. *Physical chemistry: a molecular approach*. Sausalito, CA: University Science Books, 1997. 1,270 p. 0935702997 QD453.2 541 97-142

Builds up physical chemistry, from considerations of atoms to systems containing numerous molecules. The first half is devoted to quantum mechanics and its applications to chemical problems. The second half covers more-traditional physical chemistry topics such as gases, the laws of thermodynamics, and chemical kinetics. Some topics expected in a comprehensive physical chemistry book are excluded; there is no electrochemistry and little solution physical chemistry.

4.044

Mortimer, Robert G. *Physical chemistry*. 2nd ed. San Diego: Harcourt/Academic Press, 2000. 1,116 p. 0125083459 QD453.2 541.3 99-68611

Discusses systems, states, and processes; equilibrium macroscopic states of gases and liquids; work, heat, and energy: the First Law of Thermodynamics; the Second and Third Laws of Thermodynamics: entropy; thermodynamics of real systems; phase equilibrium; multicomponent systems; chemical equilibrium, constants, the principle of Le Chatelier; thermodynamics of electrochemical system electrolytes; gas kinetic theory; transport processes; reaction rates and mechanisms; quantum mechanics; the Schrödinger equation; electronic states of atoms; the hydrogen atom and simple orbital approximations for multielectron atoms; translational, rotational, and vibrational states of atoms; spectroscopy and photochemistry; equilibrium statistical mechanics; and structure of solids and liquids.

4.045

Rock, Peter A. *Chemical thermodynamics*. Mill Valley, CA: University Science Books, 1983. 548 p. 0935702121; 0198557124 QD504 541.3 82-51233

Classical thermodynamics for undergraduates. Introduces temperature and quantitative temperature scales, conservation of energy, computational of internal energy changes in a system, entropy function in limiting thermal energy flow, and statistical thermodynamics.

4.046

Whalen, James W. *Molecular thermodynamics: a statistical approach*. NY: Wiley, 1991. 381 p. (A Wiley-Interscience publication) 0471514780 QD504 541.369 90-39767

Extension of statistical thermodynamics presented to undergraduates; will provide a deeper understanding of the relationship between statistical thermodynamics and equilibrium thermodynamics.

Catalysis

4.047

Advances in catalysis and related subjects, ed. by Walter G. Frankenberg, V.I. Komarewsky, and E.K. Rideal. NY: Academic Press, 1948-1970. 21 v. ISSN 0065-2342 QD501 541.39 49-7755

Index for v. 1-16, 1948-64, in v. 17.

4.048

Catalysis and zeolites: fundamentals and applications, ed. by J. Weitkamp and L. Puppe. NY: Springer, 1999. 380 p. 3540636501 TP245.S5 660 98-48578

Comprehensively overviews the applications of zeolites in industrial catalysis; provides an understanding of how zeolites are synthesized, modified, and characterized; and explains shape-selective catalysis.

4.049

Encyclopedia of catalysis; Istvan T. Horvath, [ed. in chief]. Hoboken, NJ: Wiley-Interscience, 2003. 6 v. 0471241830 (set) QD505 660 2002-27422

Discusses the principles of various types of catalysis, including homogeneous, heterogeneous, and biological catalysis; the preparation and characterization of catalysts; the scope of catalytic reactions and use of theory and computation in catalysis; and reaction engineering and modeling of catalytic process as well as thorough desciptions of existing catalytic technologies.

4.050

Gates, Bruce C. *Catalytic chemistry.* NY: Wiley, 1992. 458 p. 0471517615 QD505 541.3 91-4192

An outstanding work on catalysis in solutions, by enzymes, in molecular cavities (including zeolites and other molecular sieves), on surfaces, and by polymers. Central ideas are chemical; principles are illustrated using industrial reactions and catalysts.

Photochemistry

4.051

Advances in photochemistry, ed. by W.A. Noyes et al. NY: Interscience-Wiley, 1963- . Publication suspended 1980-86. Irregular. ISSN 0065-3152 QD601 541.35082 63-13592

V. 26, 2001; editors: Douglas C. Neckers, Gunther Von Bunau, William S. Jenks. Throughout the past 20 years, this series explored the frontiers of photochemistry through chapters written by experts who evaluate existing data critically. Critical and authoritative evaluations of developments in every area of the discipline. Illustrations, charts, tables.

4.052

Calvert, Jack G., and James N. Pitts. *Photochemistry.* NY: Wiley, 1966. 899 p. QD601 541.35 65-24288

4.053

Kopecký, Jan. *Organic photochemistry: a visual approach.* NY: VCH, 1992. 285 p. 0895732963 (VCH Publishers); 3527269525 (VCH Verlagsgesellschaft) QD275 547.1 91-36270

Thorough and useful for beginners owing to the complete and clear explanations. Good development of basic photochemistry and basic quantum mechanics theory. Focuses primarily on organic photochemical reactions.

4.054

Organic and inorganic photochemistry, ed. by V. Ramamurthy and Kirk S. Schanze. NY: M. Dekker, 1998. 355 p. (Molecular and supramolecular photochemistry, 2) 0824701747 QD714 541.3 98-35550

Discusses electron transfer within DNA complexes; coordination complexes and nucleic acids; electron transfer and binding; electron transfer in metallorganics; liquid crystalline polymer photochemistry and photophysics; solid-to-solid photochemical reactions; chemical and photophysical processes in multiphoton excitation; and environmental photochemistry with semiconductors.

4.055

Van Hecke, Gerald, and Kerry K. Karukstis. *A guide to lasers in chemistry.* Boston: Jones and Bartlett, 1998. 252 p. 0763704121 QD63.L3 542 97-13440

Treats laser theory, classes of laser equipment, and uses of lasers as spectroscopic probes or in photochemical processes.

Quantum Chemistry

4.056

Advances in quantum chemistry, [originally] ed. by Per-Olov Löwdin. NY: Academic Press, 1964- . Irregular. ISSN 0065-3276 QD453 541.383082 64-8029

Includes articles and invited reviews by leading international researchers. Quantum chemistry involves the treatment of the electronic structure of atoms, molecules, and crystalline matter, and describes it in terms of electron wave patterns. Uses physical and chemical insight, sophisticated mathematics, and computers to solve the wave equation and produce results.

4.057

Atkins, P.W., and R.S. Friedman. *Molecular quantum mechanics.* 3rd ed. Oxford; NY: Oxford University Press, 1997. 545 p. 0198559488; 019855947x pbk QD462 541.2 96-23892

This work ranges from a review of junior-level physical chemistry with an emphasis on nomenclature and notation, to electric and magnetic properties of matter, through the new chapter on scattering theory. An understanding of symmetry, as well as matrix algebra and calculus, is assumed.

4.058

Baggott, J.E. *The meaning of quantum theory: a guide for students of chemistry and physics.* Oxford; NY: Oxford University Press, 1992. 230 p. (Oxford science publications) 0198555768; 019855575x pbk QC174.12 530.1 91-34937

Introduces the quantum practitioner to fundamental philosophies and critical phenomenological tests. Guides readers through Einstein-Podolsky-Rosen thought experiments, Bell's inequality, and tests demonstrating the non-local nature of quantum mechanical measurements.

4.059

Bates, David R. *Quantum theory.* NY: Academic Press, 1961-1962. 3 v. (Pure and applied physics, v. 10) QC174.1 530.12 59-15762

V. 1: Elements; v. 2: Aggregates of particles; v. 3: Radiation and high energy physics.

4.060

Clark, Tim, and Rainer Koch. *The chemist's electronic book of orbitals.* Berlin; NY: Springer, 1999. 96 p. 1 CD-ROM 3540637265 QD461 541.2 98-46052

Introduces orbitals using 3D and VRML representation of molecular orbitals. The CD-ROM offers a short introduction to basic chemistry and physics of orbitals and demonstrates how to use VRML techniques. The disc contains an interactive textbook and a selection of classical organic compounds and inorganic complex ligands (including their orbitals) that may be interactively altered to study the behavior of orbitals.

4.061

Dykstra, Clifford E. *Quantum chemistry and molecular spectroscopy.* Englewood Cliffs, NJ: Prentice Hall, 1992. 470 p. 0137473125 QD462 541.2 91-25868

Bibliographic references included. Unconventional order used: quantum mechanics of molecules comes before atoms; matrix mechanics and the Schrödinger equation approach are both used, and classical mechanics is introduced where appropriate. Less rigorous treatment of derivations of complete wave functions for the harmonic oscillator, rigid rotor, and hydrogen atom.

4.062

Fitts, Donald D. *Principles of quantum mechanics as applied to chemistry and chemical physics.* NY: Cambridge University Press, 1999. 351 p. 0521651247; 0521658411 pbk QD462 541.2 98-39486

Takes an intermediate approach, using series solutions for some systems and solving others, such as the harmonic oscillator, with more up-to-date operator methods. The superposition of waves and of wave packets provides the background for Schrödinger wave mechanics introduced later. Other chapters treat one-, two-, and three-dimensional systems as well as approximation methods, systems of particles, and diatomic molecules. The point of view is formal and rigorous, with some examples from chemistry and chemical physics.

4.063

Grunwald, Ernest. *Thermodynamics of molecular species.* NY: Wiley, 1997. 323 p. (A Wiley-Interscience publication) 0471012548 QD504 541.3 96-14103

Makes thermodynamics appropriate for studying molecular species as distinct from components, and brings classical thermodynamics into the modern world of measurement of fast times and new species. Examines transition state theory, dilute solution chemistry, and equilibrium displacement measurements; linear free-energy relationships like Hammett's acidity function, discussions of entropy-enthalpy compensation, and problems with data acquired in hydrogen-bonding solutions. The approach builds on composition trees, a set of theorems bearing simple titles (correspondence, tolerance, stability), and proofs.

Other Special Topics

4.064

Borghi, Roland, and Michel Destriau. *Combustion and flames: chemical and physical principles,* with the collaboration of Gérard de Soete; transl. by Richard Turner. Paris: Éditions Technip, 1998. 371 p. 2710807408 QD516 541.3 99-208655

Introduces the scientific study of combustion and flame phenomena for new entrants to the field. This short study moves from excellent descriptive sections with very little math, to theoretical discussions that call for a fair understanding of both mathematical principles and physical and chemical science. Very good illustrations, together with worked examples of mathematical problems, make a very complicated subject easier to comprehend. Offers a nonmathematical introduction to flame structure, general combustion concepts, and discussions of engine and burner design.

4.065

Chemical kinetics and mechanism, ed. by Michael Mortimer and Peter Taylor. Cambridge, UK: Royal Society of Chemistry, 2002. 262 p. 1 CD-ROM (The molecular world) 0854046704 QD502 541.394

Presents a systematic approach to the chemistry of the p-block elements and hydrogen, and chemical bonding (oxidation numbers, bond strengths, dipole moments, and intermolecular forces). Discusses the biological role of nitric oxide and hydrogen bonding, and the new chemistry of carbon nanotubes.

4.066

Franklin, Joseph Louis. *Ion-molecule reactions*, ed. by J.L. Franklin. Stroudsburg, PA: Dowden, Hutchinson & Ross; NY: distributed worldwide by Academic Press, 1979. 2 v. (Benchmark papers in physical chemistry and chemical physics, 3) 0879333316 (pt. 1) QD501 541 78-16358

Contents: v. 1: Kinetics and dynamics; v. 2: Elevated pressures and long reaction times. Collection of 82 original research articles on gas phase ion-molecule reactions. Articles are grouped according to contents, and each group is preceded by explanations by Franklin.

4.067

Glassman, Irvin. *Combustion*. 3rd ed. San Diego, CA: Academic Press, 1996. 631 p. 0122858522 QD516 541.3 96-3069

A well-written volume having very useful appendixes with thermochemical, kinetic, and bond energy data, and a list of computer programs for combustion kinetics, together with

problems and copious references for further reading at the end of each chapter. An excellent addition to any reference collection in the field of combustion. It treats a wide range of subjects pertinent to the field of combustion, but calls for a background in thermodynamics and kinetics.

4.068

Jordan, Peter C. *Chemical kinetics and transport*. NY: Plenum Press, 1979. 368 p. 0306401223 QD501 541 78-20999

A well-written, practically error-free, accessible treatise on laser photochemistry, oscillating reactions, and molecular beam studies used in transport phenomena examination.

4.069

Levine, Ira N. *Quantum chemistry*. 5th ed. Upper Saddle River, NJ: Prentice Hall, 2000. 739 p. 0136855121; 0136855113 pbk QD462 541.2 99-28558

Discusses the Schrödinger equation, particle in a box, operators, harmonic oscillator, angular momentum, the hydrogen atom, theorems of quantum mechanics, variation method, perturbation theory, electron spin and the Pauli principle, many-electron atoms, molecular symmetry, electronic structure of diatomic molecules, virial theorem and Hellmann-Feynman theorem, ab initio and density-functional treatments of molecules, semiempirical and molecular-mechanics treatments of molecules, and comparison of methods.

4.070

Masel, Richard I. *Chemical kinetics and catalysis*. NY: Wiley-Interscience, 2001. 952 p. 0471241970 QD502 541.3 00-43303

The 14 chapters (11 on kinetics, 3 on catalysis) discuss reaction mechanisms, solvents as catalysts, the Woodward-Hoffmann rules, and linear free-energy relationships. Other topics, such as the history of kinetic studies and activation barriers in reactions, are thoroughly developed.

4.071

Simons, Jack, and Jeff Nichols. *Quantum mechanics in chemistry*. NY: Oxford University Press, 1997. 612 p. (Topics in physical chemistry) 0195082001 QD462 541.2 96-34013

One of the most comprehensive works on this topic, but

one that does require an understanding of matrix algebra and calculus, as well as molecular symmetry. A passing familiarity with computers and programming, using freely available Fortran programs for quantitative determination of electronic structures of molecules, is needed. Coverage ranges from basic quantum mechanics and molecular orbital theory, through spectroscopy and quantum mechanical treatment of molecular collisions, to computational methods. The last section provides a good basis for understanding molecular calculation programs.

4.072

Tsukerblat, Boris S. *Group theory in chemistry and spectroscopy: a simple guide to advanced usage.* London; San Diego: Academic Press: Harcourt Brace, 1994. 430 p. (Theoretical chemistry) 0127022856 QD455.3.G75 540.15122

Chapters cover symmetry operations and groups, point groups, and classes using many molecules as examples. A general understanding of the fundamentals of symmetry is assumed. Matrix theory and vectors are used to derive group representations and character tables. Group theory is related to crystal field theory and to directed valence and molecular orbital theories. Other chapters apply groups to topics such as optical spectra, double groups in spin-orbit interactions, electron paramagnetic resonance, and metal cluster compounds; and relate group theory to vibrational spectra and electron-vibrational interactions.

PHYSICOCHEMICAL CALCULATIONS

4.073

Daubert, Thomas E., and Ronald P. Danner. *Data compilation tables of properties of pure compounds.* NY: Design Institute for Physical Property Data, American Institute of Chemical Engineers, 1985. Various pagings; loose-leaf 0816903417 TP200 660.2 85-71969

Contains data on viscosity, vapor pressure, physical constants, thermal conductivity, thermodynamic properties, surface tension, critical constants, dipole moment, heat capacity, heat of dissociation, heat of fusion, and heat of vaporization.

4.074

Malinowski, Edmund R. *Factor analysis in chemistry.*

3rd ed. NY: Wiley, 2002. 414 p. 0471134791 QD39.3.F33 542 2001-46660

The topic is both important and complex; this complexity is reflected in the book, which is liberally sprinkled with matrix algebra. If one has an infrared spectrum of the atmosphere, there are thousands of observable peaks, most from normal components, some from pollutants. For many such spectra, from which researchers wish to determine the quantities of normal atmospheric constituents and what pollutants are observed, this book is the bible on the subject. The book is well written, well researched, and up-to-date.

4.075

Methods in computational chemistry, ed. by Stephen Wilson. NY: Plenum, 1987-<1992 > v. <1-5 > 0306426455 (v. 1) QD39.3 542 87-7249

Computer disk (5.25") in pocket of v. 2. V. 1: Electron correlation in atoms and molecules. V. 2: Relativistic effects in atoms and molecules. V. 3: Concurrent computations in chemical calculations. V. 4: Molecular vibrations. V. 5: Atomic and molecular properties.

4.076

Quantities, units, and symbols in physical chemistry, prepd. for publication by Ian Mills et al. 2nd ed. Oxford; Boston: Blackwell; Boca Raton, FL: CRC Press (distributor), 1993. 166 p. (IUPAC publication) 0632035722; 0632035838 pbk QD451.5 541.3 92-40104

The 1988 ed. was rev. ed. of *Manual of Symbols and Terminology for Physicochemical Quantities and Units*, 1979, from IUPAC. Treats physical chemistry as a whole and covers aspects of terminology, symbols, and units of physical quantities.

PHYSICOCHEMICAL PROPERTIES AND DATA COMPILATIONS

Physical Property Data

4.077

American Institute of Physics. *Temperature, its measurement and control in science and industry;* papers presented at a symposium held in New York City, November 1939, under the auspices of the American

Institute of Physics, with the cooperation of National Bureau of Standards, National Research Council, and officers and committees of American Ceramic Society, American Chemical Society, American Institute of Mining and Metallurgical Engineers [and others]. NY: Reinhold, 1941- < >. 4 v. in 8. ISSN 0091-9322 QC271 536.5 41-3959

Bibliography at end of most of the papers. 4th Symposium (Columbus, OH, March 1961) is v.3, ed. by Charles M. Herzfeld; NY: Reinhold, 1962. Pt.1: Basic concepts, standards and methods; Pt.2: Applied methods and instruments; Pt.3: Biology and medicine. V. 6 publ. in 1992.

4.078
Baum, Edward J. *Chemical property estimation: theory and application.* Boca Raton, FL: Lewis Publishers, 1998. 386 p. 0873719387 TD193 628.5 97-33124

Describes modern methods of estimating chemical properties that are more efficient than traditional techniques. Discusses accuracy and reliability of the methods; treats partitioning of chemicals in the environment; includes regression methods and group contribution methods; and collects and organizes current information from environmental, industrial, and computational chemical research.

4.079
Dreisbach, Robert Rickert. *Physical properties of chemical compounds.* Washington: American Chemical Society, 1955-61. 3 v. (Advances in Chemistry Series, no. 15, 22, 29) QD1 547.083 55-2887

4.080
Firestone, Richard B. *Table of isotopes*; Richard B. Firestone; Coral M. Baglin, editor; S.Y. Frank Chu, CD-ROM editor. 8th ed, 1999 update with CD-ROM. NY: Wiley, 1999. 218 p. (A Wiley-Interscience publication) 0471356336 QD601.2 541.38840212 99-25447

Compiles nuclear structure and decay data for more than 2,100 isotopes and isomers. Search software is included on the CD-ROM disc.

4.081
Jordan, Thomas Earl. *Vapor pressures of organic compounds.* NY: Interscience, 1954. 266 p. QD476 536.44 52-11414

4.082
Kaye, G. W. C., and T.H. Laby. *Tables of physical and chemical constants and some mathematical functions.* 16th ed. Essex, UK; NY: Longman, 1995. 611 p. 0582226295 QC61 530 95-15665

4.083
National Research Council, Office of Critical Tables. *Consolidated index of selected property values: physical chemistry and thermodynamics.* Washington, DC: National Academy of Sciences—National Research Council, 1962. 274 p. (National Research Council Publication no. 976) QD65 541.36083 62-60077

4.084
National Standard Reference Data System (NSRDS): conceived as a continuance of the International Critical Tables in 1963 — but not as a single publication or volume. It is essentially numerous monographic pamphlets in single subjects available for nominal fees from NSRDS. A list of the available items is found at the back of the latest editions of the *CRC Handbook of Chemistry and Physics* under the title "Numerical Data Projects."

4.085
Smithsonian physical tables, 9th rev. ed. ed. by W.E. Forsythe. Washington, DC: Smithsonian Institution, 1959. 827 pp. 4th rev. reprint, 1969. (Smithsonian miscellaneous collections, v. 120) Q11 530.8 54-60067

Azeotropes and Solutions

4.086
Azeotropic data, Jürgen Gmehling ... [et al.]. Weinheim; NY: VCH, 1994. 2 v., 1,793 p. 3527286713 QD526 660.28425 95-115329

Contains 8,800 tables of azeotropes used in separation techniques; 22 figures.

4.087
Azeotropic data, compiled by Lee H. Horsley and coworkers at the Dow Chemical Co. Washington, DC: American Chemical Society, 1952-1973. 3 v. (Advances in chemistry series, no. 6, 35, 116). 0841201668 QD1 541.36 52-3085

V. 1 subtitle: Tables of azeotropes and nonazeotropes; v. 2 compiled by L.H. Horsley and William S. Tamplin. Contains tables of azeotropes and non-azeotropes; vapor-liquid equilibrium for more than 17,000 systems; tables of binary, ternary, quaternary, and quinary systems. Includes formula index, bibliography, and charts for calculating composition and predicting azeotropism.

4.088

Burgess, John. *Ions in solution: basic principles of chemical interactions*. 2nd ed. Chichester, England: Horwood, 1999. 222 p. 1898563500 QD561 541.372

(1st ed., 1988.) A concise introduction to behavior and properties of ions in aqueous solutions. Treats solvation, thermodynamics, and kinetics of inorganic solution systems. Extensive bibliography.

4.089

Butler, James N., with David R. Cogley. *Ionic equilibrium: solubility and pH calculations*. NY: Wiley, 1998. 559 p. (A Wiley-Interscience publication) 0471585262 QD561 541.3 97-13435

Describes how to calculate exact or approximate concentration values for simple and complex ionic solutions. Includes discussions of how to use a spreadsheet program on a personal computer to perform iterative calculations and solve for equilibrium concentrations in complex systems. The computational methods developed range from those appropriate for simple equilibrium expressions of monoprotic acids and bases to multiple equilibria in complex media like seawater and geologic samples. Early discussion of activity coefficients and a presentation of software packages with sample calculations. Sample problems and exercises.

4.090

Fogg, Peter G.T., and William Gerrard. *Solubility of gases in liquids: a critical evaluation of gas/liquid systems in theory and practice*. Chichester; NY: Wiley, 1991. 332 p. 0471929255 QD543 541.3 90-45670

Includes data on common laboratory and industrial gases and solvents. Tables of data at atmospheric pressure and at a variety of temperatures; very current references. Includes smoothing equations for gas solubilities for computer graphics presentations and interpolation of data.

4.091

Gas-phase ion and neutral thermochemistry, by Sharon G. Lias [et al.] NY: Published by American Chemical Society and the American Institute of Physics for the National Bureau of Standards, 1988. 861 p. (Journal of Physical and Chemical Reference Data: v. 17, 1988, Supplement no. 1) 0883185628 QC702 88-70606

4.092

Gmehling, Jürgen, U. Onken, and W. Arlt. *Vapor-liquid equilibrium data collection*. Frankfurt/Main: Dechema; Flushing, NY: Distributed by Scholium International, 1977-<1996 >. v. 1, 1a-b, 2a-f, 3-4, 3a-b, 4a, 5, 6a-c, 8, in 18 v. (Chemistry data series, vol. 1) 3921567092 (v. 2) QD503 541.3 79-670289

Pt. 1: Aqueous-organic systems; Pt. 1a: Aqueous-organic systems (supplement 1); Pt. 1b: Aqueous systems (supplement 2); Pt. 2a: Organic hydroxy compounds, alcohols; Pt. 2b: Organic hydroxy compounds, alcohols and phenols; Pt. 2c: Organic hydroxy compounds, alcohols (supplement 1); Pt. 2d: Organic hydroxy compounds, alcohols and phenols (supplement 2); Pt. 2e: Organic hydroxy compounds, alcohols (supplement 3); Pt. 2f: Organic hydroxy compounds, alcohols and phenols (supplement 4); Pts. 3-4: Aldehydes and ketones, ethers; Pt. 3a: Aldehydes (supplement 1); Pt. 3b: Ketones (supplement 1); Pt. 4a: Ethers (supplement 1); Pt. 5: Carboxylic acids, anhydrides, esters; Pt. 6a: Aliphatic hydrocarbons, C4-C6; Pt. 6b: Aliphatic hydrocarbons, C7-C18; Pt. 6c: Aliphatic hydrocarbons (supplement 1); Pt. 8: Halogen, nitrogen, sulfur, and other compounds.

4.093

Hansen, Charles M. *Hansen solubility parameters: a user's handbook*. Boca Raton, FL: CRC Press, 2000. 208 p. 0849315255 QD543 547 99-26234

Solubility parameters, once used primarily in the coatings industry, are now valuable in the areas where polymer interfaces are critical. Chapters offer relevant background and significance of solubility parameters and the equations used to determine them; describe methods for characterizing surfaces, pigments and fillers, and filled polymer systems; and describe the application of solubility parameters to barrier polymers, coatings and filled polymer systems, and the environment. Concludes with chapters on the Hansen solubility parameters of biological materials and on future applications, and an appendix with a table of solubility parameters of numerous solvents.

4.094

Hartley, F.R., C. Burgess, and R.M. Alcock. *Solution

Physical Chemistr

equilibria. Chichester, UK: W. Horwood; NY: Halsted Press, 1980. 361 p. 0470268808 (Halsted Press) QD503 541.3 79-42956

Authoritative work that provides theoretical and experimental methods for the study of solution equilibria and determination of associated stability constants.

4.095

Hildebrand, Joel H., John Prausnitz, and Robert L. Scott. *Regular and related solutions.* NY: Van Nostrand Reinhold, 1970. 228 p. QD541 541 79-122670

Revised version of *Regular Solutions*, 1962, by Joel H. Hildebrand and Robert L. Scott. Contents: The regular solution concept; Thermodynamic relations; The liquid state; Intermolecular forces; Entropy of mixing; Heat of mixing; Gibbs free energies of liquid mixtures; Solutions of gases; Solutions of solids; Liquid-liquid solubility. Appendixes contain symbols, solubilities, and other critical data.

4.096

Hillert, Mats. *Phase equilibria, phase diagrams, and phase transformations: their thermodynamic basis.* Cambridge, UK; NY: Cambridge University Press, 1998. 538 p. 0521562708; 0521565847 pbk QD503 541.3 97-12280

Describes theoretical and practical principles of phase diagrams and application in materials science, emphasizing the importance of modern computer calculations to predict and estimate the phase diagram information from basic thermodynamic databases. Discusses principles of thermodynamics, description of chemical composition, and a general evaluation of the driving force or stability in a system under given conditions. Phase diagrams are developed from the corresponding Gibbs free energy diagrams. Discussion also includes chemical equilibria calculations and phase transformation reactions, which are separated into sharp or gradual phase transformations, transformations at constant composition, and partitionless transformations. The last few chapters describe various types and methods of mathematical modeling of phase diagrams and physical properties of the system under consideration.

4.097

Horvath, Ari L. *Handbook of aqueous electrolyte solutions: physical properties, estimation, and correlation methods.* NY: Ellis Horwood; Halsted Press, 1985. 631 p. (Ellis Horwood series in physical chemistry) 0853128944; 0470202149 (Halsted Press) QD565 541.3 85-14039

Data on electrolyte solutions, with indexes and bibliography.

4.098

Linke, William F. *Solubilities, inorganic and metal organic compounds; a compilation of solubility data from the periodical literature.* 4th ed. Washington: American Chemical Soc., 1958-65. 2 v. QD66 541.34 65-6490

A revision and continuation of the compilation by Atherton Seidell. Arranged alphabetically by element symbol. Literature citations are included.

4.099

Marcus, Y. *The properties of solvents.* Chichester; NY: Wiley, 1998. 239 p. (Wiley series in solution chemistry) 0471983691 QD544 541.3 98-18212

Provides, for some 250 solvents, physical and chemical properties, purification, toxicity, reaction rates, electrical and optical properties, surface and transport properties, polarity, acidity and basicity, partition functions, spectroscopic and electrochemical properties, and other pertinent information.

4.100

Murrell, J.N., and A.D. Jenkins. *Properties of liquids and solutions,* 2nd. ed. Chichester; NY: Wiley, 1994. 303 p. 0471944181; 047194419X pbk QD541 540 93-046721

Broad coverage of the chemical aspects of the liquid state. Covers electrolytes, nonelectrolytes, colloids, and liquid crystals. Contents: Liquid state, intermolecular forces, theories and models of the liquid state, thermodynamic properties of pure liquids, liquid crystals: the mesophase, mixtures of nonelectrolytes, phase diagrams for multicomponent systems, polar liquids, aqueous solutions of electrolytes, chemical equilibria in solutions, solutions of polymers, liquid interfaces and adsorption phenomena, colloidal systems. (New edition of 1982 work.)

4.101

Poling, Bruce E., Robert C. Reid, and John M. Prausnitz. *The properties of gases and liquids.* 5th ed. NY: McGraw-Hill, 2001. 1 v., various pagings 0070116822 TP242 660 00-61622

Revision of 4th ed., 1987. Contents: Estimation of physical properties; Pure component constants; PVT relations of pure gases and liquids; Volumetric properties of mixtures; Thermodynamic properties of ideal gases; Vapor pressures and enthalpies of vapor of pure liquids; Fluid phase equilibria in multicomponent systems; Viscosity; Thermal conductivity; Diffusion coefficients; Surface Tension. Appendixes contain property data, Lennard-Jones potentials.

4.102

Seidell, Atherton. *Solubilities of inorganic and metal organic compounds; a compilation of quantitative solubility data from the periodical literature.* 3rd ed. NY: Van Nostrand, 1940-1941. 2 v. QD66 541.342 40-34917

2nd ed. was titled *Solubilities of Inorganic and Organic Compounds....* Both volumes have Chemical Abstracts references to data. Elements are alphabetically arranged by symbol, and solubilities in mixtures of solvents are included. V. 2: Solubilities of organic compounds.

4.103

Sillén, Lars Gunnar. *Stability constants of metal-ion complexes.* 2nd ed. London: The Chemical Society, 1964. 754 p. (Special publication 17) QD503 541.3 78-238341

Cover title: *Stability Constants.* First ed., 1957-58, by Jannik Bjerrum, G. Schwarzenbach, and L.G. Sillén was published as the Chemical Society's Special publication no. 6-7. Section I: Inorganic ligands, by L. G. Sillén; Section II: Organic ligands, by A. E. Martell.

4.104

Solubilities of inorganic and organic compounds, ed. by H. Stephen and T. Stephen. Oxford: Pergamon; NY: Macmillan, 1963-1979. 3 v. in 7. 0080235999 (set) QD543 541.3 63-12770

Translation of *Spravochnik po rastvorimosti.* V. 3 ed. by Howard L. Silcock. V. 1, pt. 1-2, Binary systems; v. 2, pt. 1, Ternary systems; v. 2, pt. 2, Ternary and multicomponent systems. No critical evaluation of data.

4.105

Solubility data series. Oxford; NY: Pergamon, 1979- . Irregular. Indexes to Vols. 1 (1979)-18 (1985), as v. 19; vols. 20 (1985)-38 (1989), as v. 39. ISSN 0191-5622 QD543 541.3 85-641351

Covers a wide range of inorganic materials, including alkali metal chlorides, hydrogen halides, and precious metal halates. New volumes discuss gas solubilities, carbon monoxide, solubilities in molten salts. Each volume includes system, CASRNs, and author indexes.

4.106

Timmermans, Jean. *The physico-chemical constants of binary systems in concentrated solutions.* NY: Interscience, 1959-1960. 4 v. QD453 547.134 59-8839

V. 1: Two organic compounds (without hydroxyl derivatives); v. 2: Two organic compounds (at least one a hydroxyl derivative); v. 3: Systems with metallic compounds; v. 4: Systems with inorganic + organic or inorganic compounds (excepting metallic derivatives). Bibliography: v.4, p. 1157-1326.

4.107

Young, David A. *Phase diagrams of the elements.* Berkeley: University of California Press, 1991. 291 p. 0520074831 QD503 541.3 90-25978

Contains up-to-date phase diagrams for all the elements in the Periodic Table, and summaries of theory, experimental techniques, and results. Discussions are on an element-by-element basis.

The Chemical Bond

4.108

The chemical bond: structure and dynamics, ed. by Ahmed Zewail. Boston: Academic Press, 1992. 313 p. 0127796207 QD461 541.2 91-29643

An outgrowth of Linus Pauling's 90th birthday symposium, held at CalTech in 1991. Two chapters were contributed by Pauling himself; other contributors are among the most prominent scientists of the 20th century.

4.109

Pauling, Linus C. *The nature of the chemical bond, and the structure of molecules and crystals; an introduction to modern structural chemistry.* 3rd ed. Ithaca, NY: Cornell, 1960. 644 p. (The George Fisher Baker non-resident lectureship in chemistry at Cornell University, 18) QD469 541.396 60-16025

4.110

Stranges, Anthony N. *Electrons and valence: development of the theory, 1900-1925.* College Station: Texas A&M University Press, 1982. 291 p. 0890961247 QD469 541.2 81-48378

Authoritative and detailed account of the electronic theory

of valence from its beginnings through the discovery of electron, up to the discovery of the nuclear atom.

Colloids, Surfaces, and Interfaces

4.111
Adamson, Arthur W., and Alice P. Gast. *Physical chemistry of surfaces*, 6th ed. NY: Wiley, 1997. (A Wiley-Interscience publication) 0471148733 QD506 541.3 97-5929

Full update with more than 35% new material; treats surface chemistry from theory and models to analysis and solid-gas interfaces. Includes expanded treatment of biotechnological devices, emulsions, polymers, and new information on microscopy.

4.112
Adsorption: theory, modeling, and analysis, ed. by József Tóth. NY: Marcel Dekker, 2002. 878 p. (Surfactant science series, v. 107) 0824707478 QD547 541.3 2002-67806

Shows how to calculate mono- or multilayer adsorption, the structure of an adsorbed layer, the kinetics of each process, changes in surface free energy, and exact isotherm equations. Discusses the latest advances in rare-gas adsorption, "supercritical region" isotherms, hydrophobic solid-water interfaces, irreversible particle adsorption, and adsorption surface complexation.

4.113
Encyclopedia of surface and colloid science, ed. by Arthur T. Hubbard. NY: Marcel Dekker, 2002. 4 v., 5,667 p. 0824706331 (set) QD506

V. 1: A-Dif; v. 2: Dif-Int; v.3: Inv-Pol; v. 4: Por-Z. Collects fundamental theories and recent research from biology, biochemistry, physics, and other applied areas. Discusses the latest developments in drug delivery, enzyme behavior, and protein adsorption; current utilization of colloids, emulsions, foams, gels, and fine particles; chemical behavior at electrochemical, fluid-fluid, polymer, soil, mineral, and atmospheric interfaces; mechanisms of catalysis, transport, and adsorption; characterization of fatty acids and lipids; commercial uses for membranes; and contemporary modes of environmental remediation.

4.114
Evans, D. Fennell and Håkan Wennerström. *The colloidal domain: where physics, chemistry, biology, and technology meet*. 2nd ed. NY: Wiley-VCH, 1999. 632 p. (Advances in interfacial engineering series) 0471242470 QD549 541.3 98-23227

Discusses colloidal interactions and phenomena from interdisciplinary and practical viewpoints, with a focus on association colloids.

4.115
Handbook of surface and colloid chemistry, ed. by K.S. Birdi. 2nd ed. Boca Raton, FL: CRC Press, 2003. 765 p. 0849310792 QD508 541.3 2002-25923

In 16 chapters, provides a rigorous and broad treatment of interphase interactions in addition to appendixes of physical properties pertinent to colloid and surface interactions. Offers a critical and detailed approach to this field—empirically, theoretically, and mathematically. Also discusses spectroscopic absorption and scattering techniques used to study particle and aggregate sizes. Explains scanning probe/scanning tunneling microscopy/atomic force microscopy in the study of biopolymer structures. Extensive bibliography.

4.116
Hiemenz, Paul C., and Raj Rajagopalan. *Principles of colloid and surface chemistry*. 3rd ed., rev. and expanded. NY: Marcel Dekker, 1997. 650 p. 0824793978 QD549 541.3 97-4015

Includes new material on fractal dimensions of aggregates, intraparticle and interparticle structure, dynamic light scattering and the relation between light scattering and X-ray and neutron scattering; the Krieger-Dougherty relation; and the non-Newtonian rheological behavior of dispersion.

4.117
McCash, Elaine M. *Surface chemistry*. Oxford; NY: Oxford University Press, 2001. 177 p. 0198503288 pbk QD506 541.3 2001-270027

Introduces surface structure, bonding, the experiments and theory of absorption and desorption, surface reaction mechanisms (including heterogeneous atmospheric chemistry), and preparation and properties of ultrathin films. The first part introduces surfaces and the language of surfaces, including defining variables that are used in later chapters.

4.118

Myers, Drew. *Surfaces, interfaces, and colloids: principles and applications*. 2nd ed. NY: Wiley-VCH, 1999. 501 p. 0471330604 QD506 541.3 98-38906

Expanded edition emphasizes emulsions, foams, colloidal properties, aerosols, polymers at interfaces, membrane phenomena, solubilization, wetting, lubrication, and adhesion. Mathematics are minimally used. Useful for both industrial chemists and students. (1st ed., 1991).

4.119

Stein, H.N. *The preparation of dispersions in liquids*. NY: M. Dekker, 1996. 245 p. (Surfactant science series, v. 58) 0824796748 QD549 541.3 95-40714

Colloidal interaction and, especially, the preparation of dispersions in liquids are key parameters in many industrial processes. Stein offers a very useful guide to the scientific background of those processes. Four chapters treat basic notation and definitions, dispersion of a solid in a liquid, liquid/liquid dispersions, and foams.

4.120

Surface and colloid chemistry handbook: CRCnetBASE 1999, ed. by K.S. Birdi. CRCnetBASE, 1999. ISSN 1523-3138 QD508 541 99-3181

Application of surface and colloidal science ranges through soaps and detergents, emulsion technology, oil recovery, pollution control, food processing, foams, biomembranes, pharmaceuticals, synthetic transplants, and biological monitors. The subject matter has been effectively organized into 19 critical areas, appropriately illustrated, readily accessible, and easily converted to print using computer technology. This work has regular updating and access to the CRC database.

Crystallography

4.121

Borchardt-Ott, Walter. *Crystallography*. Berlin; NY: Springer-Verlag, 1993. 303 p. 3540566791 (Berlin); 0287566791 (NY) QD905.2 548 93-44956

Introduces fundamental of geometric crystallography. Discusses states of matter, space lattices, crystal morphology, symmetry, point groups, space groups, crystal growth, and chemical and physical properties arising from symmetry and order. Crystal structures of metals, noble

gases, ionic salts, covalent compounds, and inorganic complexes are explained.

4.122

Crystallization technology handbook, ed. by A. Mersmann. 2nd ed. NY: Marcel Dekker, 2001. 840 p. 0824705289 TP156.C7 660 2001-28553

1st ed., 1995. Treats fundamentals of crystallization; interaction between balances, processes, and product quality; design of crystallizers and crystallization processes; control of crystallizers; reaction crystallization; additives and impurities; crystallization from the melt; thermal analysis and economics of processes. Appendixes with physical processes and examples of industrial large-scale crystallizers.

4.123

Giacovazzo, Carmelo. *Direct phasing in crystallography: fundamentals and applications*. [Chester, England]: International Union of Crystallography; Oxford; NY: Oxford University Press, 1998. 767 p. (IUCr monographs on crystallography, 8) 0198500726 QD921 548 99-193578

Excellent, up-to-date coverage of a subject central to the rapidly growing field of crystal structure determinations using X-ray-, neutron-, and electron-diffraction methods. Direct phasing, the most commonly used method for extracting the positions of atoms in a crystal from experimental data. In this best comprehensive work on the subject, the author begins with first principles and presents the various methods and applications of direct phasing in a systematic, rigorous, and thorough manner. Special chapters on direct methods in macromolecular crystallography and powder diffraction data.

4.124

Glusker, Jenny Pickworth, with Mitchell Lewis and Miriam Rossi. *Crystal structure analysis for chemists and biologists*. NY: VCH, 1994 854 p. (Methods in stereochemical analysis) 0895732734 QD945 548 92-7886

Written to acquaint readers with general principles of crystal-structure analysis. The 18 authoritative chapters (basic crystallography; instrumentation; data collection, analysis and refinement; presentation and interpretation of structural results; and crystallographic databases),each have many figures and tables, summary, glossary, and extensive list of references to primary and secondary literature. Good subject separate index; index of terms defined in the glossaries. An excellent introductory text; a useful reference;

a thorough introduction to the advanced literature; and a source of comprehensive reviews of specific structure-oriented topics.

4.125

Hammond, Christopher. *The basics of crystallography and diffraction.* 2nd ed. NY: Oxford University Press, 2001. 331 p. (International Union of Crystallography texts on crystallography, 5) 0198505531; 0198505523 pbk QD905.2 548 00-53086

Covers the fundamentals of crystallography—the study of patterns, symmetry, lattices, diffraction—but does not discuss crystal-structure determination. The book's language is British English, as reflected in the choice of words and in recommended sources for obtaining models and computer programs. Fine bibliography, biographical notes on founders of crystallography, good index, many excellent figures and diagrams.

4.126

Ladd, Mark. *Crystal structures: lattices and solids in stereoview.* Chichester: Horwood, 1999. 171 p. (Horwood series in chemical science) 1898563632 QD921 548 00-690733

Introduces lattices and symmetry concepts along with descriptions of ionic, covalent, metallic, and molecular structures. The many three-dimensional figures (stereo views) do not come with a viewer. Figures are generally clear; the Periodic Table has incorrect names above atomic number 103. However, this book is packed with information; it refers to a Web site from which some computer programs (of use in solving the problems) can be downloaded; and, most importantly, as its title states, it presents a thorough elementary theoretical and descriptive discussion of the arrangements of atoms in many types of crystals.

4.127

Mak, Thomas C.W., and Gong-Du Zhou. *Crystallography in modern chemistry: a resource book of crystal structures.* NY: Wiley, 1992. 1,323 p. (A Wiley-Interscience publication) 0471547026 QD945 548 91-23395

Parallel title with author statement in Chinese. Provides a concise history of X-rays and crystallography. Three-dimensional drawings, some in stereo, of many compounds as well as cell parameters and atomic coordinates. Includes simple metallic crystals, organics, organometallics, and inclusion compounds.

4.128

Massa, Werner. *Crystal structure determination*; transl. into English by Robert O. Gould. Berlin; NY: Springer, 2000. 206 p. 3540659706 pbk QD945 548 99-36317

Crystal structure determination (CSD) allows crystallographers to locate positions of individual atoms in a crystal. Huge quantities of data, collected on computer-controlled X-ray instruments, are interpreted using standardized computer programs. This book describes the theory and practice of this process, and covers nearly every topic encountered in routine and not-so-routine CSD as well as the theories of crystal symmetry and X-ray diffraction instrumentation; the practical aspects of data collection and analysis; and advanced topics, possible errors and pitfalls, and how to interpret the results, along with worked-out CSD example with annotated computer printouts. Good index and bibliography; excellent detailed figures.

4.129

Structure and bonding in crystals, ed. by Michael O'Keeffe, Alexandra Navrotsky; contributors, Aaron N. Bloch ... [et al.]. NY: Academic Press, 1981. 2 v. 0125251017 (v. 1); 0125251025 (v. 2) QD478 548 81-7924

Chapters written by experts treat molecular-orbital theory, bond lengths, atomic and ionic radii, phase transitions, and structure types.

4.130

Woolfson, Michael M., and Fan Hai-fu. *Physical and non-physical methods of solving crystal structures.* Cambridge; NY: Cambridge University Press, 1995. 276 p. 0521412994 QD945 548 94-8254

A well-written and useful resource, with complete descriptions of methods for solving crystal structures, for crystallographers (both small molecule and macromolecular) not familiar with theoretical bases of these diverse methods. Offers a concise introduction to theory of X-ray diffraction followed by a description of theory and mathematics and practice of different structure solution methods. Chapter references. Describes Patterson and heavy-atom methods; the many strategies (and computer programs) available for using direct methods in solving a structure. Isomorphous replacement and anomalous scattering are carefully described. Many novel methods are described; the last chapter discusses multiple beam scattering methods for direct phase determination.

Electrochemistry

4.131

Bard, Allen J. *Encyclopedia of electrochemistry of the elements.* NY: M. Dekker, [1973]- 1994. v. <1-8, 9A-B, 10-15 > in <16 >. 082476093x QD551 541 73-88796

V. 11-<15 > have special title: Organic section; ed. by A.J. Bard and H. Lund.

4.132

Bard, Allen J., and Larry R. Faulkner. *Electrochemical methods: fundamentals and applications.* 2nd ed. NY: John Wiley, 2001. 833 p.

1st ed. published in 1980. Consistent, complete, and integrated material, rigorous and encyclopedic treatment of modern electrochemical theory and practice.

4.133

Bockris, John O'M., and Amulya K.N. Reddy. *Modern electrochemistry.* 2nd ed. NY: Plenum Press, 1998-2000. 3 v., 2,053 p. 0306465973 (set); 0306465981 (set) pbk; 0306455544 (v. 1); 0306455552 (v. 1) pbk; 0306461676 (v. 2A) pbk; 0306461668 (v. 2A); 0306463253 (v. 2B) pbk; 0306463245 (v. 2B) QD553 541.3 97-24151

V. 1: Ionics; v. 2A: Fundamentals of electrodics; v. 2B: Electrodics in chemistry, engineering, biology, and environmental science. V. 2A by John O'M. Bockris, Amulya K.N. Reddy, and Maria Gamboa-Aldeco; v. 2A-2B have imprint: NY: Kluwer Academic/Plenum Publishers. First ed. published in 1970. Not applied analytical procedures such as polarography and cyclic voltammetry, but current thought on theory of all electrochemical phenomena. Ionics is the study of physical electrochemistry of solutions, primarily aqueous. Electrodics is the physical chemistry of heterogeneous phenomena observed at the solution-electrode interface.

4.134

Conway, Bruce E. *Electrochemical data.* NY: Elsevier, 1952; reprinted, Greenwood, 1969. 374 p. 0837116309 QD560 541 69-10078

Contains physical property and electrochemical data on inorganic and organic compounds; includes dielectric constants, activity coefficients, electrolytic conductance, dissociation constants of acids and bases in aqueous and organic solvents, solubilities, properties of colloids and biological electrolytes, molten salts, electrode potentials, and electrode kinetics. Subject index.

4.135

Crow, David Richard. *Principles and applications of electrochemistry.* 4th ed. Cheltenham, UK: S. Thornes, 1998. 282 p. 0748743782 QD553 541.3 00-304342

Originally published: London; NY: Chapman and Hall, 1974. Contains current, basic theory that underlies the practices of electrochemistry.

4.136

Dell, Ronald M., and David A.J. Rand. *Understanding batteries.* Cambridge: Royal Society of Chemistry, 2001. 223 p. (RSC paperbacks) 0854046054 TK2896 621.31242

Provides a brief history of the development of batteries and discusses applications and markets. Discusses basic battery terminology and science; rechargeable versus nonrechargeable; and the matching of battery technology to intended applications.

4.137

The encyclopedia of electrochemistry, ed. by Clifford A. Hampel. NY, Reinhold; London, Chapman & Hall, [1964]. Reprinted by R.E. Krieger Pub. Co., 1972. 1,206 p. QD553 541 64-22288

Dictionary-style reference; 412 articles discuss a range of topics.

4.138

Hamann, Carl H., Andrew Hamnett, and Wolf Vielstich. *Electrochemistry.* Weinheim; NY: Wiley-VCH, 1998. 423 p. 3527290966; 3527290958 pbk QD553 541.3 98-155076

Electrochemistry includes the conductivity of salt solutions, the nature of electron transfer from charged surfaces, the structure of solutions near such charged surfaces, adsorption, catalysis of reactions by surfaces, mass transport (by a variety of mechanisms), chemical reaction cascades induced by electrodes, quantitative and qualitative analysis of mixtures, and so forth. This authoritative monograph covers physical electrochemistry, i.e., the properties of charged surfaces and electrolyte solutions, quite well.

4.139

Latimer, Wendell Mitchell. *The oxidation states of the elements and their potentials in aqueous solutions.* (usually known as Oxidation Potentials). 2nd ed. NY:

Prentice-Hall, 1964. 392 p. (Prentice-Hall chemistry series) QD561 52-10791

2nd ed., 1952. Contents: Units, conventions and general methods; Ionization potentials, electron affinities, lattice energies, and relationship to standard oxidation-reduction potentials; Elements.

4.140

Nonaqueous electrochemistry, ed. by Doron Aurbach. NY: Marcel Dekker, 1999. 602 p. 0824773349 QD555.5 541.3 99-35915

Electrochemistry includes corrosion, batteries, and indeed all reactions involving chemical oxidation or reduction, and touches on a variety of important areas such as chemical analysis, synthesis, absorption, and nonaqueous solvents, including not only organic media but (more recently) new media such as solid polymer electrolytes and molten salts. This is a major area of industrial research, e.g., in the search for high energy, low-pollution batteries. This timely book is the most complete treatment of nonaqueous organic solvent systems since the pioneering work by Charles K. Mann and Karen Barnes, *Electrochemical Reactions in Nonaqueous Systems* (1970).

4.141

Sawyer, Donald T., Andrzej Sobkowiak, and Julian L. Roberts. *Electrochemistry for chemists*. 2nd ed. NY: Wiley, 1995. 505 p. (A Wiley-Interscience publication) 0471594687 QD553 541.3 95-2738

(Rev. ed. of *Experimental Electrochemistry for Chemists*, 1974.) Modern electrochemistry is practiced by physical chemists to obtain fundamental thermodynamic and kinetic data, analytical chemists interested in new theoretical and experimental techniques, organic chemists preparing novel materials, and other specialists (biochemists, engineers) to obtain yet other types of information. A wide variety of techniques are described, with further references to the original literature. The coverage is broad but not very deep. A number of sections cover the descriptive electrochemistry of a number of common systems.

Thermochemical and Thermophysical Data

4.142

Barin, Ihsan. *Thermochemical data of pure substances*, in collaboration with Fried Sauert, Ernst Schultze-Rhonhof,

and Wang Shu Sheng. 2nd ed. Weinheim, FRG; NY: VCH, 1993. 2 v., 1,739 p. 1560817178 (NY; set); 3527285318 (Weinheim; set) QD511.8 541.3 92-46891

V. 1: Ag-Kr; v. 2: La-Zr. Lists basic thermochemical data for about 2,400 chemicals, arranged alphabetically by chemical formula. For each chemical entry, nine basic constants are provided for 100-degree intervals above room temperature.

4.143

Baulch, D. L., J. Durbury, et al. *Evaluated kinetic data for high temperature reactions*. London: Butterworths, 1972-81. 4 v. 0408703466 QD502 541 76-371748

V. 1: Homogeneous gas phase reactions of the H_2-O_2 system. v. 3: Homogeneous gas phase reactions of the O_2-O_3 system, the CO-O2-H2 system, and of sulphur-containing species. V. 4 published as *Journal of Physical and Chemical Reference Data* v. 10, 1981, Supplement 1. (JPCRD-NBS; 10).

4.144

Binnewies, M., and E. Milke. *Thermochemical data of elements and compounds*. Weinheim; NY: Wiley-VCH, 1999. 928 p. 3527297758 QD511.8 541.3 99-190836

Contains selected values of thermochemical properties of more than 3000 mostly inorganic substances. Values are given for enthalpy of formation and entropy at 298K, as well as enthalpy, entropy, and Gibbs free energies at other temperatures via a polynomial expression.

4.145

Chase, M. W. Jr. [et al.] *JANAF thermochemical tables*. 3rd ed. Washington, DC: American Chemical Society; NY: American Institute of Physics for the National Bureau of Standards; 1986. 2 v. (Journal of Physical and Chemical Reference Data v. 14, 1985; Supplement 1) 0883184737 QD511.8 530.8 85-71387

4.146

Domalski, Eugene S., William H. Evans, and Elizabeth D. Hearing. *Heat capacities and entropies of organic compounds in the condensed phase*. Washington, DC: American Chemical Society; NY: American Institute of Physics for National Bureau of Standards, 1984. 286 p. (Journal of Physical and Chemical Reference Data, v.13, Supplement 1) 0883184478 QC145.4 530.4 84-70909

Contains a 22-page bibliography.

4.147

Dreisbach, Robert Rickert, compiler. *Pressure-volume-temperature relationship of organic compounds; a reference volume for reading directly the variation of vapor pressure with temperature of a compound belonging to one of the twenty-three Cox chart families ...* 3rd ed. Sandusky, OH: Handbook Publishers, 1952. 303 p. QD518 541.36 52-2687

Published in 1944 as *Table of Vapor Pressure-temperature Charts.* Useful for finding boiling points at various pressures. Compiles pressure-temperature and latent heat data.

4.148

Gammon, Bruce E., Kenneth N. Marsh, and Ashok K.R. Dewan. *Transport properties and related thermodynamic data of binary mixtures.* 2 v. NY: Design Institute of Physical Property Data (DIPPR), American Institute of Chemical Engineers, 1993- . 0816905800 (pt. 1); 0816906220 (pt. 2) QD541 661.8 93-33434

Each volume covers transport properties (viscosity, thermal conductivity, and diffusion coefficients), liquid property data (solubility in various solvents, excess volume, mixture density, and surface tension), and process design data (mixture critical properties).

4.149

Ho, C. Y., R.W. Powell, and P.E. Liley. *Thermal conductivity of the elements: a comprehensive review.* Washington, DC: American Chemical Society, 1975. 796 p. (Journal of Physical and Chemical Reference Data 3, 1974, Supplement 1) 0883182165 QD466 541 75-4440

4.150

Hultgren, Ralph R., et al. *Selected values of the thermodynamic properties of the elements.* Metals Park, OH: American Society of Metals, 1973. 636 p. QD171 546 73-76587

4.151

[Hultgren] Selected values of thermodynamic properties of binary alloys, prepared by Ralph R. Hultgren [and others]. Metals Park, OH: American Society for Metals, 1973. 1,435 p. QD171 546 73-76588

These two volumes together represent a complete revision of the 1963 work *Selected Values of Thermodynamic Properties of Metals and Alloys* by Hultgren, Orr, Anderson, and Kelley.

4.152

IUPAC. Commission on Thermodynamics and Thermochemistry. *Experimental thermochemistry: measurement of heats of reaction,* ed. by Frederick D. Rossini and Henry A. Skinner. NY: Interscience, 1956, 1962. 2 v. QD511 541.36 55-11450

V. 1 edited by Frederick D. Rossini; v. 2 edited by H.A. Skinner.

4.153

IUPAC, Physical Chemistry Division. *Experimental chemical thermodynamics, v. 1: Combustion calorimetry,* ed. by Stig Sunner and Margaret Mansson. Oxford; NY: Pergamon Press, 1978- . 0080209238 (v. 1) QD511 541 77-30694

Topics include units and physical constants, basic principles of combustion calorimetry, reduction of data to standard states, assignment of uncertainties, and combustion of compounds with nonmetallic heteroatoms, metals and metallic compounds, organometallics, combustion in fluorine, flame calorimetry, microcalorimetry, and history of combustion calorimetry.

4.154

NIST-JANAF thermochemical tables; Malcolm W. Chase, Jr. 4th ed. [Washington, DC:] American Chemical Society; Woodbury, NY: American Institute of Physics for the National Institute of Standards and Technology, 1998. 2 v., 1,952 p. (Journal of physical and chemical reference data. Monograph no. 9) 1563968312 (set); 1563968193 (pt. 1); 1563968207 (pt. 2) QD511 541.3 98-86732

Pt. 1: Al-Co; Pt. 2: Cr-Zn. Contents: History of the JANAF thermochemical tables; notation and terminology; reference states and conversions; evaluation of thermodynamic data; construction of the tables; additions, revisions, and corrections; references; indexes to the tables; the tables themselves, arranged as in chemical formula index.

4.155

Physicochemical properties and environmental fate handbook [computer file], [Donald Mackay, Wan-Ying Shiu, and Kuo-Ching Ma]. Boca Raton, FL: Chapman & Hall/CRCnetBase, 1999. 1 CD-ROM 084939757x TD196.O73 628.5 98-12469

The handbooks contained on this disc include physical-chemical data for similarly structured groups of chemical substances that influence their fate in air, water, soils, sediments, and their resident biota. Includes environmental laws and legislation; environmental impact assessment; prevention, treatment, and management of air pollution; wastewater management; groundwater and stormwater pollutant management; solid waste management, including incineration; hazardous waste disposal and management; toxic and radioactive waste-site management.

4.156

Rossini, Frederick D. [and others]. *Selected values of chemical thermodynamic properties*. Washington, DC: U.S. Government Printing Office, 1952. 1,268 p. (National Bureau of Standards Circular 500) QC100 541.36083 52-60677

4.157

Rossini, Frederick D., et al. *Selected values of physical and thermodynamic properties of hydrocarbons and related compounds*, comprising the tables of the American Petroleum Institute Research Project 44 extant as of December 31, 1952. Pittsburgh: American Petroleum Institute by Carnegie Press, 1953. 1,050 p.

Bibliography: p. 972-1039. "Publication of the API Research Project 44": p. 1048-1050. Also: API Project 44: Tables of selected values of properties of hydrocarbons and related compounds, NY: API, 1964. The title is self-explanatory, but it is essential that the introduction be read before using the tables.

4.158

Stull, D.R., and Harold Prophet. *JANAF thermodynamical tables*. Sponsored by the Advanced Research Projects Agency (Washington DC) with the assistance of the Joint-Army-Navy-Air Forces. 2nd ed. Washington, DC: U.S. Government Printing Office, 1971. 1 v., unpaged. (NSRDS-NBS-37) QD511

Prepared at Thermal Research Laboratory, Dow Chemical Co., Midland, MI. Supersedes PB 168370 and addenda 1,2, and 3. Tables of thermodynamic properties arranged by chemical symbol.

4.159

Thermal constants of substances; editor, V.S. Yungman; founding editors, V.P. Glushko, V.A. Medvedev, L.V. Gurvich. NY: Begell House; John Wiley, 1999. 8 v. in 10 parts. 0471318558 TA418.52 620.1 98-29638

V. 1, pts. I-III; v. 2: pt. IV; v. 3: pt. V; v. 4, pt. VI; v. 5: pt. VII; v. 6: pt. VIII; v. 7, pt. IX; v. 8: pt. X. This work contains about 26,000 substances and more than 51,500 cited references (some as far back as the 1800s) on self-consistent thermal constants of all the inorganic, simple organic, and metallorganic substances. Included are enthalpy and Gibbs free energy of formation, dissociation energy, enthalpy content, entropy and heat capacity at standard temperature, temperatures and enthalpies of phase transition, crystallographic and critical parameters, ionization potential, and electron affinity.

4.160

Thermodynamic properties of individual substances; editors, L.V. Gurvich and I.V. Veyts, C.B. Alcock. NY: Begell House, 1996- . Various volumes. 1567000754 (v. 3, pt. 1-2) QD504

USSR Academy of Sciences, Institute for High Temperatures and State Institute of Applied Chemistry in cooperation with the National Standard Reference Data Service of the USSR. Revised and updated from the third Russian edition. V. 1: Elements O, H (D,T), F, Cl, Br, I, He, Ne, Ar, Kr, Xe, Rn, S, N, P and their compounds. Pt. 1: Methods and computation; Pt. 2: Tables; v. 2: Elements C, Si, Ge, Sn, Pb, and their compounds. Pt. 1: Methods and computation. Pt. 2: Tables; v. 3: Elements B, Al, Ga, In, Tl, Be, Mg, Ca, Sr, Ba and their compounds. Pt. 1: Methods and computation. Pt. 2: Tables (2 v. in 1).

4.161

U.S. National Bureau of Standards. *Tables of chemical kinetics, homogeneous reactions,* [by] National Bureau of Standards and National Research Council, Committee on Tables of Constants. Editorial office: Frick Chemical Laboratory, Princeton University. Washington, DC: U.S. Government Printing Office, 1951. 1 v. (looseleaf). (NBS Circular 510) QC100 541.083 52-60566

4.162

Wagman, Donald D., and Frederick D. Rossini. *Selected values of chemical thermodynamic properties.* Washington, DC: U.S. National Bureau of Standards; for sale by U.S. Government Printing Office, 1965. (NBS technical note 270) QC100 541.36 65-62923

An advance issue of a revision of the tables of series I of the National Bureau of Standards Circular 500, *Selected Values of Chemical Thermodynamic Properties*, by F.D. Rossini [and others]. Pt. 1: Tables for the first twenty-three elements in the standard order of arrangement; Pt. 2: Tables for the elements twenty-three through thirty-two in the standard order of arrangement; Pt. 3: tables for the first thirty-four elements in the standard order of arrangement; Pt. 4: Tables for elements 35 through 53 in the standard order of arrangement; Pt. 5: tables for elements 54 through 61 in the standard order of arrangement.

4.163

Wilhoit, R. C., and B.J. Zwolinski. *Physical and thermodynamic properties of aliphatic alcohols*. NY: American Chemical Society, 1973. 420 p. (JPCRD-NBS v. 2, 1973, Supplement 1) 0883182025; 088318205x pbk QD305 547 73-84178

4.164

Yaws, Carl L. *Thermodynamic and physical property data*. Houston: Gulf Publ. Co., 1992. 217 p. 0884150313 QD504 547.1 91-36702

Successor to: *Physical Properties: A Guide to the Physical, Thermodynamic and Transport Property Data of Industrially Important Chemical Compounds*. NY: McGraw-Hill, 1977.

Atomic and Molecular Data

4.165

Haile, J.M. *Molecular dynamics simulation: elementary methods*. NY: Wiley, 1992. 489 p. (Physical chemistry series) 0471819662 QC168.7 532 91-31963

Fundamentals; Hard spheres; Finite-difference methods; Soft spheres; Static properties; Dynamic properties.

4.166

Howell, M. Gertrude, Andrew S. Kende, and John S. Webb, editors. *Formula index to NMR literature data*; contributors: V. Bauer [and others]. NY: Plenum, 1965- <1966 >. v. <1-2 > QD291 547 64-7756

V. 1: References prior to 1961. V. 2: 1961-1962 references. V. 1 indexes the literature for some 2,500 compounds whose NMR spectra were reported in the literature up to 1961. V. 2 is the formula index for 1961-62 references, arranged by P, S, N, O, I, Br, Cl, F, C.

4.167

McClellan, Aubrey Lester. *Tables of experimental dipole moments*. San Francisco: W.H. Freeman, 1963. 713 p. QD571 541.377 63-14844

Bibliography, p. 587-684.

4.168

Streitwieser, Andrew, Jr., and John I. Brauman. *Supplemental tables of molecular orbital calculations*. With a *Dictionary of Pi-electron Calculations*, by C.A. Coulson and A. Streitwieser Jr. Oxford; NY: Pergamon Press, 1965. 2 v. QD461 547.122 63-10069

The *Dictionary of Pi-electron Calculations* has also been issued separately. Contains tables of Hückel molecular orbital calculations of pi-systems, showing energy and coefficients of each molecular orbit, total pi-energy, electron densities, bond orders, and polarizabilities.

LABORATORY TEXTS AND MANUALS

4.169

Daniels, Farrington [and others]. *Experimental physical chemistry*. 7th ed. NY: McGraw-Hill, 1970. 669 p. QD457 541 75-77952

4.170

Salzberg, Hugh, Jack I. Morrow, and Stephen R. Cohen. *Laboratory course in physical chemistry*. NY: Academic Press, 1966. 319 p. QD457 541.3025 65-26407

4.171

Shoemaker, David P., Carl W. Garland, and Joseph W. Nibler. *Experiments in physical chemistry*. 6th ed. NY: McGraw-Hill, 1996. 778 p. 0070570744 QD457 541.3 95-44102

Contents include treatment of experimental data; gases; transport properties of gases; thermochemistry; solutions; phase equilibria; electrochemistry; chemical kinetics; surface phenomena; macromolecules; electrical and magnetic properties; spectra and molecular structure; solids; electrical measurements; temperature; vacuum technology; instruments; miscellaneous procedures; least-squares fitting procedures; use of computers.

4.172

White, John M. *Physical chemistry laboratory*

experiments. Englewood Cliffs, NJ: Prentice Hall, 1975. 563 p. 0136659276 QD457 541 74-11029

Contents include data handling and computation; experimental apparatus and methods; a collection of experiments introducing techniques; physical properties of gases; thermodynamics; kinetics; spectroscopy; bulk electric and magnetic properties; electrochemistry; macromolecular physical chemistry; molecular structure.

PROBLEM MANUALS AND PHYSICAL CHEMISTRY SOFTWARE

4.173

Cropper, William H. *Mathematica computer programs for physical chemistry.* NY: Springer, 1998. 246 p. 0387983376 pbk QD455.3E4 541.3 97-34136

Many formerly daunting computations and equations can be very effectively explored using software such as Mathematica. The accompanying CD-ROM offers programs covering more than a hundred topics commonly encountered in undergraduate and introductory graduate physical chemistry courses.

4.174

Field, Martin J. *A practical introduction to the simulation of molecular systems.* Cambridge; NY: Cambridge University Press, 1999. 325 p. 052158129x QD480 541.2 98-37540

Treats molecular modeling and will be of greatest value to readers who are willing to do some computer programming, or at least maintain a set of programs. The programs and modules of this book, known as The DYNAMO Molecular Simulation Library, are available online. The specific programming language used is Fortran 90. Emphasis is principally on approaches, which begin with molecular rather than quantum mechanics. Thus the methods discussed can be applied to fairly large systems, including proteins.

4.175

Francl, Michelle M. *Survival guide for physical chemistry.* Lakeville, MN: Physics Curriculum & Instruction, Inc., 2001. 136 p. 0971313407 QD458 2001-93987

A portable and readable compendium of information needed by the student studying physical chemistry that

emphasizes mechanics rather than theory. Examples and annotated solutions are provided.

4.176

Metz, Clyde R. *Schaum's outline of theory and problems of physical chemistry.* 2nd ed. NY: McGraw-Hill, 1989. 500 p. (Schaum's outline series) 0070417156 QD456 541.3 87-29839

Contains explanations of all principal concepts in physical chemistry, coverage of all course fundamentals, supplements to all major texts, examples, and worked problems.

4.177

P-chem [computer file], [developed by] Philip R. Watson. Wiley, 1997. 1 CD-ROM; user manual. Windows Ver. 1.0 0471162264 QD453.2

P-chem very comprehensively and interactively covers major topics in an undergraduate course in physical chemistry—thermodynamics, statistical thermodynamics, quantum chemistry, chemical reactions and kinetics, macromolecules, surface chemistry, and solid state. The program is easy to install and needs 10 MB of hard drive space. Graphics are excellent, in color and in two or three dimensions. For each module, various systems can be investigated over a range of temperatures and pressures. Data can be inserted from a list of predetermined molecules or that of the user's choice. Animation is included. A summary, explanatory text that reviews the underlying theory is provided for each module.

4.178

Whittaker, A.G., A.R. Mount, and M.R. Heal. *Physical chemistry.* Oxford: Bios Scientific; NY: Springer, 2000. 286 p. (Instant notes) 0387916199 QD453.2 541.3

This review book in outline form emphasizes readability and a logical progression through physical chemistry. Each section (i.e., chapter) starts with "Key Notes" that present important concepts; important terms and phases are in boldface, and illustrations are clear and informative with good legends. The presentation is not always rigorous, as in thermodynamics, where the normalization of thermodynamic parameters to the molar basis is not indicated (e.g., enthalpies of combustion and formation; there is a lack of stoichiometric coefficients in the equations using standard reaction enthalpies). A supplementary aid for introductory students of physical chemistry.

Chapter 5

Organic Chemistry

In this adventure through the chemical literature, the vast area of organic chemistry will be surveyed through its literature. You will note that there is more literature in organic than in other sections of chemistry, because there are simply more possibilities for atom-atom combinations of carbon and other heteroatoms, as well as of carbon-carbon itself. Carbon, as you remember, is unique among the Periodic Table members in that it is able to chain together to create many intricate and large compounds; no other atom can make that claim! (Silicon comes close, but the stability of its silicon-silicon bonds are not as great as that of carbon-carbon, so there will probably be very few, if any, silicon-based forms of life in the universe. So, save your movie money.)

Organic chemistry is one of the most important areas of study because the chemistry of life itself is represented here. This chapter examines organic chemical literature with some general material applicable to all types of compounds. A later part of this chapter treats areas of organic chemistry that are actually applications or derivations of the general topics considered in the first part. Here will be discussed such topics as natural products and organic analysis, as well as spectroscopy and stereochemistry of organic compounds. Future sections will treat environmental chemistry and biochemistry. The literature of environmental chemistry has increased dramatically during the last 25 years, encompassing toxicology, workplace safety and health, and a host of other related topics.

GUIDES TO THE LITERATURE

5.001
The Beilstein system: strategies for effective searching, ed. by Stephen R. Heller. Washington, DC: American Chemical Society, 1998. 208 p. 0841235236 QD257.7 025.06 97-35536

Contents: The Beilstein system: an introduction; The Beilstein handbook; The Beilstein online database; Current facts in chemistry on CD-ROM; Computer systems for substructure searching; CrossFireplusReactions; Using the Beilstein reaction database in an academic environment; Beilstein's CrossFire: a milestone in chemical information and interlibrary cooperation in academia; use of the Beilstein system in the chemical and pharmaceutical industries; AutoNom.

5.002
Sommerville, Arleen N. "Information sources for organic chemistry," *Journal of Chemical Education*, 68(7), 553-561, 1991; 68(10), 842-853, 1991; 69(5), 379-386, 1992.

Discusses information sources in organic chemistry, in a format suitable for classroom use. Various searching strategies are discussed, based on name reactions, functional groups, and reagents.

BIBLIOGRAPHIES, INDEXES, AND REVIEWS

5.003
ChemPrep [electronic resource]. Philadelphia, PA: Institute for Scientific Information, 1997- . CD-ROMs. ISSN 1091-8558 Quarterly updates. Back years from 1985. 96-3711

Provides a comprehensive review and retrieval of organic literature from some 100 international journals. Quarterly updates ensure current material, and each year is cumulated. Data are searchable by structure, text, keyword, or author. Users may draw structures with a special tool included; structures are searchable by reactant or product. Text and reaction structure searches may be combined via Boolean operators. Journals may be browsed via issue, reaction, or by article. Purchase price is comparable to subscription to a large corpus of journals.

5.004

ChemReact nn. Berlin; Heidelberg: Springer, 1997- . CD-ROMs.

This is a chemical reaction search system with *nn* thousand (*nn* may be at least 32; in 1997 was 41) organic reaction types from journals worldwide. Reactions were selected from the *InfoChem Reaction Database* which has about 1.8 million reactions.

5.005

Current chemical reactions. Philadelphia, PA: Institute for Scientific Information, 1979- . Print and CD-ROM; back years from 1986 on CD-ROM. ISSN 0163-6278 547 80-944

Covers synthetic methods from worldwide journals, accessing more than 400,000 reactions. Retrievable information includes complete reaction diagrams, critical conditions, bibliographic data, and author abstracts. Updated every six months.

5.006

Current facts [computer file]; (Beilstein current facts in chemistry). Berlin; NY: Springer-Verlag, 1990- . CD-ROMs. Quarterly updates. ISSN 0939-7698 QD251.B43 96-47022

This electronic product is an electronic version of *Beilstein's Handbuch der organischen Chemie*. The CD-ROM includes the most recent 12 months of organic chemistry on one disc. Each disc contains information on some 300,000 compounds with chemical and physical data, extracted from 82 international organic journals. Menus offer data and substructure searches, including stereochemical searches.

5.007

InfoChem reaction database. Berlin; Heidelberg: Springer, 1997- . CD-ROM software.

InfoChem is used to display results from a structure search in *ChemReact nn* (see above). Users may search reaction types, documents, or bibliographic information.

5.008

The Ring index, 2nd ed. Ed. by Austin McDowell Patterson, Leonard T. Capell, and D.F. Walker. Washington, DC: American Chemical Society, 1960. 1,425 p. ISSN 0080-309x (for supplement to 2nd ed.) QD291 547 41-84 (1st ed.)

(1st ed., 1940.) Covers literature through 1956; Supplement I, 1963, covers the literature 1957-1959; Supplement II, 1964, covers 1960-1961; Supplement III, 1965, covers 1962-1963. A most useful compilation of data on cyclic compounds. Any system devoid of functional groups (e.g., naphthalene, but not naphthoic acid) that has been mentioned in the literature will be included. A citation in *The Ring Index* does not guarantee that a synthesis has been achieved. Rings are classified by size, and under each such section there are subdivisions by number of ring members of each element. Appendix: Rules for naming and numbering ring systems used in organic chemistry, p. 1267-1306.

NOMENCLATURE

5.009

Fox, Robert B., and Warren H. Powell. *Nomenclature of organic compounds: principles and practice*. 2nd ed. Oxford; NY: Oxford University Press; Washington, DC: American Chemical Society, 2001. 437 p. 0841236488 QD291 547 99-43801

(Revision of *Nomenclature of Organic Compounds*, ed. by John H. Fletcher, Otis C. Dermer, and Robert B. Fox, 1974.) Discusses origin and evolution of organic nomenclature; conventions in organic nomenclature; common errors, pitfalls, and misunderstandings; acyclic, alicyclic, arene hydrocarbons; hydrocarbon ring assemblies; heteroacyclic and heterocyclic compounds; groups cited only by prefixes in substitutive nomenclature; carboxylic acids, acid halides, and replacement analogs; carboxylic esters, salts, and anhydrides; aldehydes and chalcogen analogs; ketones and chalcogen analogs; alcohols; phenols; ethers; peroxides and hydroperoxides; carboxylic amides, hydrazides, and imides; amidines and other nitrogen analogs of amides; nitriles; amines and imines; other nitrogen compounds; sulfur, selenium, and tellurium acids and their derivatives; thiols, sulfides, sulfoxides, sulfones, and their chalcogen analogs; phosphorus and arsenic compounds; silicon, germanium, tin, and lead compounds; boron compounds; organometallic compounds; polymers; stereoisomers; natural products; isotopically modified compounds; radicals, ions, and radical ions. Appendixes include prefixes, common endings, and a glossary.

5.010

Fresenius, Philipp. *Organic chemical nomenclature*, by Philipp Fresenius with collaboration from Klaus Gö. Chichester: Ellis Horwood; NY: Halsted Press 1989. 294 p.

(Ellis Horwood series in organic chemistry) 0470210982 QD291 547 88-21599

Translated by A.J. Dunsdon from the German 2nd ed. (1983) of *Organisch-chemische Nomenklature: Einfuhrung in die Grundlagen mit Regeln und Beispielen*.

5.011

Giese, Friedo. *Beilstein's index: trivial names in systematic nomenclature of organic chemistry*. Berlin; NY: Springer-Verlag, 1986. 253 p. 0387161422 QD291 547 85-27914

Contains listings of trivial names; root modifying prefixes; IUPAC rules; ring index; formula index; CASRNs.

5.012

Godly, Edward W. *Naming organic compounds: a systematic instruction manual*. 2nd ed. Hemel Hempstead, England: E. Horwood for LGC and Commission of the European Communities, 1995. 264 p. (Ellis Horwood series in organic chemistry) 0131036238 QD291 547 94-27486

A user-friendly guide with quick, systematic access to procedures for naming new compounds. Pull-out chart sends users to appropriate numbered section for instructions. Offers preferred IUPAC nomenclature.

5.013

Hellwinkel, Dieter. *Systematic nomenclature of organic chemistry: a directory to comprehension and application of its basic principles*. Berlin; NY: Springer, 2001. 228 p. 3540411380 pbk QD291 547 2001-20540

Introduces systematic organic nomenclature; offers simple and concise guidelines for generation of systematic names as codified by IUPAC rules. Besides common compounds, the book offers information on naming cyclophanes, carbohydrates, organometallics, and isotopically modified compounds. Stereochemical specifications are discussed.

5.014

IUPAC. Commission on the Nomenclature of Organic Chemistry. *Nomenclature of organic chemistry*. 4th ed. Prepared for publication by J. Rigaudy and S.P. Klesney. Oxford; NY: Pergamon Press, 1979. 559 p. 0080223699 QD291 547 79-40358

Contains 1979 ed. of Sections A, B, C, D, E, F, and H. 1st ed. issued by the International Union of Pure and Applied Chemistry. Section A: Hydrocarbons; B: Fundamental heterocyclic systems; C: Characteristic groups containing carbon, hydrogen, oxygen, nitrogen, halogens, sulfur, selenium and/or tellurium; D: Organic compounds containing elements not exclusively those in Section C; E: Stereochemistry; F: General principles for naming natural products; H: Isotopically modified compounds.

5.015

Names, synonyms, and structures of organic compounds: a CRC reference handbook, ed. by David R. Lide and G.W.A. Milne. Boca Raton, FL: CRC Press, 1995. 3 v. (2162, 266, 651 p.) 0849304059 (set) QD291 547 94-25283

V. 1, CAS number 50-00-0 to 2433-97-8; v. 2, CAS number 2435-53-2 to 127665-32-1; v. 3, Molecular formula index, name/synonym index.

5.016

University of Cincinnati. Department of Chemistry. Organic Division; Milton Orchin, Fred Kaplan, Roger S. Macomber, et al. *The vocabulary of organic chemistry*. NY: Wiley, 1980. 609 p. 0471044911 QD291 547 79-25930

Definitions and examples for more than 1,200 of the most common and most important terms in organic chemistry, arranged in 15 chapters.

ENCYCLOPEDIAS

5.017

Encyclopedia of reagents for organic synthesis; editor-in-chief, Leo A. Paquette. Chichester; NY: Wiley, 1995. 8 v. (6,223 p.) 0471936235 (set); 0471938718 QD77 547.2 9532803

Online version available as *e-EROS*. Includes nearly 48,000 reactions and about 3500 most frequently used reagents. Online version is searchable by structure, substructure, reagent, reaction type, experimental conditions, and more. Entries highlight various uses and characteristics of each reagent. Information provided on reagents includes physical data, solubility, form supplied, purification, preparation, and literature references. Electronic version follows:

e-EROS [electronic resource]: Encyclopedia of reagents for organic synthesis. NY: John Wiley & Sons, 2001- . QD77

Online version of the *Encyclopedia of Reagents for Organic Synthesis* (see previous entry). Available by subscription via the Internet.

DICTIONARY

5.018
Patai, Saul. *Glossary of organic chemistry, including physical organic chemistry*. NY: Interscience Publishers, 1962. 227 p. QD251 547.03 62-12995

This volume defines terms peculiar to organic chemistry, and includes structures and equations where appropriate. References are coded at beginning of volume.

HANDBOOKS AND TABLES OF DATA

5.019
Beilstein, Friedrich Konrad. *Handbuch der organischen Chemie*. 4. Aufl. Die Literatur bis 1. Januar 1910 umfassend, hrsg. von der Deutschen Chemischen Gesellschaft, bearb. von Bernhard Prager und Paul Jacobson, unter standiger Mitwirkung von Paul Schmidt und Dora Stern. ... Berlin; New York: Springer, 1918-40; 1941-57; 1972- . v.1-31. QD251 547 22-79

Tables; cumulative index to basic series and 1st-4th suppl: v.1, subject index (1975); v.1, formula index (1975); v.17-18, subject index (1977). Each volume will have a cumulative index. One of the world's greatest and most useful information sources for organic chemistry. Formerly all in German; newer volumes are either bilingual or in English. Some editions were available in microcard. There is an online version through major vendors (chiefly, STN). After initial publication of two volumes in five parts in 1882, after 20 years of work. The third edition, in four volumes, four supplements, and an index, was dated 1892-1906. A second edition (three volumes) was published in 1886-1890. The original 27 volumes constitute the Haupwerk. Fifteen supplementary volumes make up the first supplement (erstes Ergänzungswerk), for literature to 1920. A second edition (zweites Ergänzungswerk) was begun in 1941 to cover literature from 1920-1929; a third/fourth edition was initiated (drittes und viertes Ergänzungswerk) in 1974 to cover literature from 1930-1959; and a fourth edition (viertes Ergänzungswerk) in 1972 to cover literature from 1950-1972. Not all volumes were updated in these newer editions.

5.020
Beyer, Hans and Wolfgang Walter. *Handbook of organic chemistry*, transl. by Douglas Lloyd. London; NY: Prentice Hall, 1996. 1,037 p. 013010356x QD251.2 547 93-11632

(Transl. and amended from the 22nd German ed.) Includes many functional groups and their reactions, and provides references to comprehensive reviews of the chemistry of these groups. Mechanisms are not ignored but are given much less emphasis than in most US works. Spectral methods are briefly reviewed and frequently referred to, but interpretation of spectra is not considered. The topics isoprenoids, heterocyclic compounds, nucleic acids, enzymes, and metabolic processes occupy the last third of the book.

5.021
Bovey, Frank Alden. *NMR data tables for organic compounds*. NY: Interscience, 1967- . QC762 547 67-20258

Includes bibliography.

5.022
CRC handbook series in organic electrochemistry. Louis Meites, Petr Zuman, and William J. Scott [et al]. Cleveland, OH: CRC Press, 1977-<1983 >. v. <1-6 > (Handbook series in organic electrochemistry) 0849372208 (set) QD272 547 77-24273

V. 6- by Louis Meites, Petr Zuman, Elinore B. Rupp [et al]; v. 5-<6- > published in Boca Raton, FL. Covers literature of 1960-71. Large set of electrochemical data on organics and organometallics, arranged by formula.

5.023
Dictionary of organic compounds. 6th ed. London; NY: Chapman & Hall, 1996-1998. v. <1-9, 11 >. 0412540908 (set); 0412541203 (second supplement) QD246 547 94-69994

Kept up-to-date by annual supplements. User guide in English and Japanese. This dictionary contains more than 150,000 compounds, with CASRN, references, basic property data, and is searchable using structure, name, formula, and other data. Online version is *Heilbron*.

5.024
Dictionary of organic compounds on CD-ROM. London: Chapman & Hall Chemical Database, 1993- .

Semiannual updates. Began with Version 1.1, March 1993. ISSN 0967-6686 QD246 94-37154

This is an evaluated database of organic substances, containing chemical, physical, and biological data together with key literature references and structure diagrams. This is the CD-ROM version of the latest print edition of the *Dictionary of Organic Compounds* and all supplements, *Dictionary of Organophosphorus Compounds*, and some other Chapman & Hall publications except the natural products included in *Dictionary of Natural Products* (CD-ROM).

5.025

Dictionary of organometallic compounds. Editors: Jane E. Macintyre and J. Buckingham. London; NY: Chapman and Hall, 1984-1990. 3 v.; 5 supplements; structure index to suppl. 1-5. 0412247100 (set) QD411 547 84-19952

Describes structural, chemical, physical, and biological properties of some 30,000 organometallic compounds considered to be important.

5.026

Egloff, Gustav. *Physical constants of hydrocarbons*. NY: Reinhold, 1939-1953. 5 v. (American Chemical Society. Monograph series, no. 78) QD291 547.2 39-7977

Includes bibliographies. V 1: Paraffins, olefins, acetylenes, and other aliphatic hydrocarbons; v. 2: Cyclanes, cyclenes, cyclynes, and other alicyclic hydrocarbons; v. 3: Mononuclear aromatic hydrocarbons; v. 4: Polynuclear aromatic hydrocarbons; v. 5: Paraffins, olefins, acetylenes, and other aliphatic hydrocarbons (Revised values).

5.027

Gokel, George W. *Dean's handbook of organic chemistry*. 2nd ed. NY: McGraw-Hill, 2004. 1 v, various pagings. 0071375937 QD251.3 547 2003-277466

Tabular compilation of numerical information. Brief introduction to organic nomenclature. Basic tables; Beilstein references for several thousand organic compounds; physical properties, thermodynamic properties; spectroscopic data; other applied organic chemical information. Indexes.

5.028

Handbook of data on common organic compounds, ed. by David R. Lide and G.W.A. Milne. Boca Raton, FL:

CRC Press, 1995. 3 v. (2,229, 432 p.) 0849304040 (set) QD257.7 547 94-39019

V. 1: Compounds A-E; v. 2: Compounds F-Z; v. 3: Indexes. Provides physical property data, spectral data, and chemical structures for about 12,000 common organics. Indexed by CASRN, molecular formula, and name/synonym. Information on each compound includes name, synonyms, CAS and Beilstein Registry numbers, Beilstein reference number, Merck number, line formula, molecular formula, molecular weight, melting and boiling points, vapor pressure, density, refractive index, specific rotation, color, solubility, and mass, IR, UV, and Raman spectral data.

5.029

Handbook of data on organic compounds, ed. by David Lide and G.W.A. Milne. 3rd ed. Boca Raton, FL: CRC Press, 1994. 7 v. 0849304458 (set) QD257.7 547 93-403342

V. 1: Compounds 1-5,000, Aba-Ben; v. 2: Compounds 5,001-10,000, Ben-Cha; v. 3: Compounds 10,001-15,600, Cha-Hex; v. 4: Compounds 15,601-21,599, Hex-Pho; v. 5: Compounds 21,600-27,580, Pho-Zir; v. 6: CAS Registry Number index, molecular formula index; v. 7: Name/synonym index. Also known as HODOC III. Includes Beilstein reference numbers, chemical structures, mass spectra, Merck number, linear and structural formulas, physical data, names and synonyms (see above entry for type of information included).

5.030

Handbook of data on organic compounds (HODOC), ed. by Robert C. Weast and Jeanette G. Grasselli. 2nd ed. Boca Raton, FL: CRC Press, 1989- . v. <1-11 > 0849304202 (set) QD257.7 547 88-7485

Rev. ed. of *CRC Handbook of Data on Organic Compounds*, 1985. V. 10: 1990 update, ed. by Jeanette G. Grasselli; v.11: 1991 update, ed. by Jeanette G. Grasselli and David R. Lide. HODOC, 2nd ed. (HODOC II) is the first series of yearly updates that will provide new data to keep HODOC II the definitive reference source in organic chemistry. All of the data for each compound appear in one location under the HODOC Number for that compound; includes CASRNs.

5.031

Handbook of reagents for organic synthesis. Chichester; NY: Wiley, 1999. 4 v. 0471979244 (v. 1);

0471979260 (v. 2); 0471979252 (v. 3); 0471979279 (v. 4) QD77 547 98-53088

Founding editor, Leo Paquette. V. 1: Reagents, auxiliaries, and catalysts for C-C bond formation, ed. by Robert M. Coates and Scott E. Denmark; v. 2: Oxidizing and reducing agents, ed. by Steven D. Burke and Rick L. Danheiser; v. 3: Acidic and basic reagents, ed. by Hans J. Reich and James H. Rigby; v. 4: Activating agents and protecting groups, ed. by Anthony J. Pearson and William R. Roush. Provides detailed information topics selected from the more extensive (and more expensive) eight-volume *Encyclopedia of Reagents for Organic Synthesis* (EROS) (1995). V. 1 details the use of 203 reagents, divided into 22 classes based on chemical structure and are arranged alphabetically. The addition of a "Related Reagents" section is especially valuable in that it allows the reader to locate relevant material both in the other volumes and in the original EROS. V. 2 outlines 145 of the most important oxidizing and reducing agents, and includes a detailed discussion of related synthetic transformations effected by these agents not necessarily involved in the oxidation or reduction itself. New material (not found in the original EROS) is included. V. 3 discusses traditional acids and bases and complexing agents, ligands, and biocatalysts. V. 4 contains an alphabetical listing of both well-known and novel reagents. The decision on which reagents to include was based on describing only those reagents "that are used stoichiometrically, and that are relatively familiar to the organic chemistry community." Contains references to relevant articles from *Organic Synthesis,* review articles, and monographs. Lists of general abbreviations and reference abbreviations, contributors, reagent formula index, and extensive subject index. Clear structural representations; concise, well-written descriptions of the reagents; amazingly error-free text.

5.032

Lide, David R. *Handbook of organic solvents.* Boca Raton, FL: CRC Press, 1995. 565 p. 0849389305 TP247.5 661.8 94-42775

Offers information on some 500 organics used as solvents. Includes information on health hazards, safety guidelines, limiting values for air exposure, carcinogenicity, and other hazard ratings. Data include refractive index; spectral data fields; vapor pressures; heats of fusion, formation, combustion, and vaporization; acid-base dissociation constants, and more. Indexes include CASRN, melting and boiling points, name/synonym, and molecular formulas.

5.033

Rappoport, Zvi, compiler. *CRC handbook of tables for organic compound identification.* 3rd ed. Cleveland, OH: Chemical Rubber Co., [1967]. 564 p. QD291 547 63-19660

Some 8100 parent compounds are arranged by functional group, with subarrangements in order of increasing melting point, boiling point, etc. Indexes.

5.034

Reagents for organic synthesis (Fieser and Fieser's reagents for organic synthesis). Ed. by Mary Fieser and others. v. 1-7, NY: Wiley, [1967-72]; v. 8, 1980- . 7 v.; v. 1-12 in 1 v. V. 10 is 1983. Irregular. ISSN 0271-616x QD262 547.2 88-659800

Original editors, Louis F. Fieser and Mary Fieser. V. 1-12 available in one volume, including index to earlier volumes. V. 1, 1967 (1,457 p.), had index to reagents, A-Z, suppliers, index of apparatus and of reagent by type, and subject and author indexes. For each reagent: structure, molecular weight, melting and boiling points, preparation, other reactions, and references. Older title: *Reagents for Organic Synthesis*, by Louis Frederick Fieser and Mary Fieser, 1967-1986. Authors' names in reverse order on v. 2-7; v. 8, <10-12 > by Mary Fieser. V. 9 by Mary Fieser, Rick L. Danheiser, William Roush. V. 8-<12 > have title: *Fieser and Fieser's Reagents for Organic Synthesis.*

5.035

Riddick, John A., William B. Bunger, and Theodore K. Sakano. *Organic solvents: physical properties and methods of purification.* 4th ed. NY: Wiley, 1986. 1,325 p. (Techniques of chemistry, v.2) 0471084670 QD61 542 86-15698

Reference source for physical properties of solvents and purification methods. Physical property data needed by organic chemists are found in the first half of the book; purification methods are described in detail, along with references to important journal articles, in the second part. For both sections, solvents are grouped by type of compound, such as halogenated hydrocarbons, esters, and hydroxy compounds. Therefore, the fastest way to locate a substance is to check the index. Values calculated by the authors are designated by underlining the references from which the basic data were taken. Additional references of interest are also included. Beilstein references and spectroscopic data are also provided. Extensive references to journal articles are provided, noted with the number in brackets. ("On the basis of the 1st ed. by A. Weissberger and

E.S. Proskauer; the completely revised 2nd ed. by John A. Riddick and Emory E. Toops, Jr.; and the completely revised 3rd ed. by John A. Riddick and William B. Bunger.")

5.036

Science of synthesis: Houben-Weyl methods of molecular transformations. Stuttgart; NY: Thieme, 2000- . 3131121319 (v. 1); 0865779406 (v. 1) QD262 547 00-61560

Successor to *Methoden der organischen Chemie (Houben-Weyl)*, 4th ed. (see above). Category 1. Organometallics. V. 1: Compounds with transition metal-carbon [pi]-bonds and compounds of groups 10-8 (ni, Pd, Pt, Co, Rh, Ir, Fe, Ru, Os). V. 4: Compounds of group 15 and silicon compounds. Category 2. Hetarenes and related ring systems. V. 9: Fully unsaturated small-ring heterocycles and monocyclic five-membered hetarenes with one heteroatom. V. 10: Fused five-membered hetarenes with one heteroatom. V. 11: Five-membered hetarenes with one chalcogen and one additional heteroatom. Table of contents [categories 1-2]. Available also via the Internet.

5.037

Structure determination of organic compounds: tables of spectral data, by Ernö Pretsch, Philippe Bühlmann, and Christian Affolter. 3rd completely rev. and enl. English ed. Berlin; NY: Springer, 2000. 421 p.; 1 CD-ROM. 3540678158 QC462.85 547 00-63719

Revised ed. of *Tables of Spectral Data for Structure Determination of Organic Compounds*, 2nd. ed., 1989. Based on the German *Tabellen zur Strukturaufklärung organischer Verbindungen mit spektroskopischen Methoden.* Includes 13C NMR, 1H NMR, IR, mass, and UV/Visible spectral data. Includes CD-ROM with software for estimating NMR chemical shifts and generating isomers based on structural information.

5.038

Ueno, Keihei, Toshiaki Imamura, and K.L. Cheng. *Handbook of organic analytical reagents.* 2nd ed. Boca Raton, FL: CRC Press, 1992. 615 p. 0849342872 QD77 547.3 91-46306

Revised ed. of *CRC Handbook of Organic Analytical Reagents*, ed. by K.L. Cheng, Keihei Ueno, and Toshiaki Imamura, 1982. Includes information on physicochemical properties, preparation, and analytical applications of the most common organic reagents. Some 40 new reagents

have been added to this new edition, a reagent index is new in this edition, and references are completely updated. Entries include synonyms, sources and methods of synthesis, analytical applications, complexation reactions, properties of complexes, purification, purity of the reagent, and other reagents with related structures.

5.039

Utermark, Walther and Walter Schicke. *Schmelzpunkt-Tabellen organischer verbindungen/ Melting point tables of organic compounds*. 2nd rev. and suppl. ed. NY: Interscience, 1963. 715 p. QD518 547.1362 63-5577

Contains tables of melting points of organic compounds, including structures-catchword index in French, Russian, English, German—gives Beilstein reference and main physical constants.

HISTORY

5.040

Ball, Philip. *Designing the molecular world: chemistry at the frontier*. Princeton, NJ: Princeton University Press, 1994. 376 p. 0691000581 QD31.2 540 93-38151

Ball includes the discovery of the so-called "buckyball" molecule, surface and enzyme catalysis, laser-induced chemical reactions, quasi-crystals, self-replicating molecules, organic polymers that conduct electricity, chemical evolution, fractals and chaos theory in chemistry, and atmospheric chemistry.

5.041

Benfey, Otto Theodor. *From vital force to structural formulas: the development of ideas in organic chemistry*. Philadelphia, PA: Beckman Center for the History of Chemistry, 1992. 115 p. (Publication/ Beckman Center for the History of Chemistry, no. 10) 0941901092 QD476 60-19555

Now back in print, this teaching tool traces the transformation of organic chemistry from 1800 to Couper and Kekulé, and notes gaps in their structural theories that were filled in by later chemists. (First published in 1964, first reprinted 1975.)

5.042

Essays on the history of organic chemistry, ed. by

Organic Chemistry

James G. Traynham. Baton Rouge, LA: Louisiana State University Press, 1987. 145 p. 0807112933 QD248 547 86-14377

Based on papers delivered at the Sixteenth LSU Mardi Gras Symposium in Organic Chemistry, 1984.

5.043

Hopf, Henning. *Classics in hydrocarbon chemistry: syntheses, concepts, perspectives*. Weinheim; NY: Wiley-VCH, 2000. 547 p. 3527302166; 3527296069 pbk QD305.H5 547 00-710434

Emphasizes throughout the synthesis and properties of "unnatural" products, materials not found on Earth except inside a lab. Treats cage hydrocarbons, prismanes, allenes, cumulenes, adamantane, annulenes, and so forth.

5.044

Tarbell, Dean Stanley, and Ann Tracy Tarbell. *Essays on the history of organic chemistry in the United States, 1875-1955*. Nashville, TN: Folio Publishers, 1986. 433 p. 0939454033 QD248.5 547 86-12062

A comprehensive and reliable recounting of the development of organic chemistry; organized according to sections such as natural products, stereochemistry, rearrangements, and polymerization.

COMPILATIONS AND TREATISES

5.045

Advances in physical organic chemistry. London; NY: Academic Press, 1963- . ISSN 0065-3160 QD476 547 62-22125

Reviews the quantitative study of organic compounds and their behavior, in language for general readers.

5.046

Carey, Francis A., and Richard J. Sundberg. *Advanced organic chemistry*. 3rd ed. NY: Plenum, 1990. 2v. Pt.A: Structure and mechanisms; Pt.B: Reactions and synthesis. 0306434407 (pt. A); 0306434474 (pt. A, pbk); 0306434563 (pt. B); 0306434571 (pt. B, pbk) QD251.2 547 90-6851

A comprehensive work for advanced undergraduates and graduate students, which professional chemists will use as a reference manual. Describes synthetically useful reactions and their applications in an easy-to-follow format using extensive reaction diagrams, with references to journal articles and handbooks. Provides an excellent description of such name reactions as Wittig, Diels-Alder, Claisen rearrangement, Birch reduction, and Cope rearrangement, all of which can be found in the index. Many reaction mechanisms are given. Each chapter concludes with a selected list of general references and practice questions.

5.047

Chemistry of carbon compounds: a modern comprehensive treatise, ed. by E.H. Rodd. rev. ed. NY: Elsevier, 1951-1989. QD251 547 51-14658

V. 1: Aliphatic compounds, 1951-2 (7 v.); v. 2: Alicyclic compounds, 1953 (5 v.); v. 3: Aromatic compounds, 1954, 1956 (8 v.); v. 4: Heterocyclic compounds, 1957, 1959, 1960 (11 v.); v. 5: Miscellaneous; indexes, 1962. This edition is followed by a newer Rodd's (see next entry).

5.048

Comprehensive organic chemistry: the synthesis and reactions of organic compounds. Chairman and deputy chairman of the editorial board, Sir Derek Barton and W. David Ollie. Oxford; NY: Pergamon Press, 1979. 6 v. 0080213197 (set) QD245 547 78-40502

V. 1: Stereochemistry, hydrocarbons, halo compounds, oxygen compounds; v. 2: Nitrogen compounds, carboxylic acids, phosphorus compounds; v. 3: Sulfur, selenium, silicon, boron, organometallic compounds; v. 4: Heterocyclic compounds; v. 5: Biological compounds; v. 6: Formula, subject, author, reaction, and reagent indexes.

5.049

Dyson, George Malcolm. *Manual of organic chemistry for advanced students*. London: Longmans, Green, 1950- . QD251 547 50-9518

Nomenclature; halogen compounds of hydrocarbons; alcohols, phenols, and ethers; aldehydes and ketones; ketenes and polyketides; acids and esters; terpenes and related compounds; polyalcohols, carbohydrates, and related compounds; steroids and other substances of biochemical interest.

5.050

Fieser, Louis F., and Mary Fieser. *Advanced organic*

chemistry. NY: Reinhold Co., 1961. 1,158 p. QD251 547 61-14594

Nature of organic compounds; structural types; stereochemistry; alkanes, alkenes, acetylene types, petroleum, alcohols, displacement reactions, halides, carboxylic acids, aldehydes and ketones, condensations, amines, ring formation and stability, history of the benzene problem, aromatic substitutions, aromatic hydrocarbons, nitro carbons, sulfonic acids, aryl amines, phenols, aryl halides, aromatic carboxylic acids, aromatic aldehydes and ketones, quinones and arenones, naphthalene, nonbenzenoid aromatics and pseudoaromatics, carbohydrates, lipids, proteins.

5.051

Harvey, Ronald G. *Polycyclic aromatic hydrocarbons.* NY: Wiley-VCH, 1997. 667 p. 1560816864; 0471186082 QD341.H9 547 96-19742

Comprehensive overview of properties, synthetic methods, and reactions of PAHs. The introduction discusses importance and sources, and explains PAH classification and nomenclature. Chapters treat structure, aromaticity, general synthesis, and reactions; most of the compounds discussed contain four to seven rings.

5.052

Harwood, Laurence M. *Polar rearrangements.* Oxford; NY: Oxford University Press, 1992. 96 p. (Oxford science publications) 0198556713; 0198556705 pbk QD281.R35 547.1 91-26884

Describes and discusses classic rearrangements (such as alkyl and hydride shifts in carbocations) and examples from natural products chemistry. Reprinted in 1999 with corrections.

5.053

Hine, Jack Sylvester. *Physical organic chemistry.* 2nd ed. NY: McGraw-Hill, 1962. 552 p. QD476 547.1 61-18627

Contents: Structure of organic molecules; Acids and bases; Kinetics and reaction mechanisms; Quantum correlations of reaction rates and equilibria; Acid-base catalysis; Addition of free radicals to olefins; Radical decomposition of peroxides and azo and diazonium compounds; Some reactions involving radical displacements; Stereochemistry and rearrangements of free radicals; Methylenes; Multicenter-type reactions.

5.054

Hydrocarbon chemistry, by George A. Olah and Árpád Molnár. NY: Wiley, 1995. 632 p. 047111359x QD305.H5 547 94-31442

Each of 12 chapters treats a type of hydrocarbon transformation and reviews all related basic chemistry, including reactivity, selectivity, stereochemistry, and numerous mechanistic aspects, together with practical applications.

5.055

Leonard, John, B. Lygo, and G. Proctor. *Advanced practical organic chemistry.* 2nd ed. London: Blackie; NY: Chapman & Hall, 1995. 298 p. 0751402001 QD262 547 94-72928

Brings together techniques and "tricks of the trade" of practicing organic chemists. Shows how to get started carrying out or analyzing reactions. 1st ed. contents: Keeping records of laboratory work; Equipping the laboratory and the bench; Purification and drying of solvents; Reagents—purification and handling; Gases; Vacuum pumps; Carrying out the reaction; Working up the reaction; Small scale reactions; Large scale reactions; Characterization; The chemical literature; Special procedures; "Troubleshooting"—what to do when things don't work; Specific examples of reactions; Safety.

5.056

Morrison, Robert Thornton, and Robert Neilson Boyd. *Organic chemistry.* 6th ed. Englewood cliffs, NJ: Prentice-Hall, 1992. 1,279 p. 0136436692 QD251.2 547 91-27444

Emphasizes activation energies and reaction mechanisms; new material on the relationship between structure and spectra. A fine textbook for undergraduates.

5.057

Organic molecular solids: properties and applications, ed. by William Jones. Boca Raton, FL: CRC Press, 1997. 426 p. 0849394287 TA418.9.C7 620.1 96-35232

"Supramolecular" systems are those in which smaller molecules are organized into much larger structures whose properties differ from those of their subunits. The field includes the so-called nanotubes (relatives of buckminsterfullerenes), Langmuir-Blodgett structures (thin, highly oriented and structured films, often on metal surfaces), conducting polymers, nonlinear optical crystals, and other novel solid substances. The current state of knowledge in these and several other areas is summarized for the first time in this book.

5.058

Organic peroxides, ed. by W. Ando. Chichester; NY: Wiley, 1992. 845 p. 041934380 QD305.E7. 547 92-1320

Chapters written by experts: Theoretical and general aspects of organic peroxides; Alkyl hydroperoxides; Dialkyl peroxides; Dioxiranes, three-membered ring cyclic peroxides; Endoperoxides; Diacyl peroxides; Sulfur and phosphorus peroxides; Peroxy acids and peroxy esters; Polyoxides; Transition metal peroxides and related compounds; Transition metal catalyzed oxidation: the role of peroxometal complexes; Peroxides from photosensitized oxidation of heteroatom compounds; Peroxides from ozonation; Formation of organoperoxides from superoxide; Peroxides in biological systems.

5.059

Progress in physical organic chemistry, ed. by S.G. Cohen, A. Streitwieser, and R.W. Taft. NY: Wiley, 1963- . (An Interscience publication) ISSN 0079-6662 QD476 547.1 63-19364

Each volume contains several chapters on different subdivisions of physical organic chemistry, written by experts. Each volume covers subjects different from others. V. 18 was published in 1990.

5.060

Rodd's Chemistry of carbon compounds; a modern comprehensive treatise. 2nd ed. Ed. by Samuel Coffey. Amsterdam; New York: Elsevier, 1964-89. v. 1, pts. A-G; v. 2, pts. A-E; v. 3, pts A-H; v. 4, pts. A-L, in 31 v. 0444406646 (set) QD251 547 64-4605

Rev. ed. of *Chemistry of Carbon Compounds*, ed. by E.H. Rodd. V.1: General introduction; aliphatic compounds (7 v.); v. 2: Alicyclic compounds (5 v.); v. 3: Aromatic compounds (8 v.); v. 4: Heterocyclic compounds (11 v.). Includes index and bibliographical references. Supplements since 1991.

5.061

Sainsbury, Malcolm. *Aromatic chemistry*. Oxford; NY: Oxford University Press, 1992. 92 p. (Oxford chemistry primers; 4) 0198556756; 0198556748 pbk QD331 547 91-47606

Summarizes fundamentals of aromatic chemistry, definition

of aromatic character and structure of benzene, reactions of arenes, electrophilic aromatic substitution, nucleophilic aromatic substitution, substituted benzenes, polycyclic arenes, and annulenes. Reprinted with corrections, 1996.

5.062

Smith, Michael, and Jerry March. *March's advanced organic chemistry: reactions, mechanisms, and structure.* 5th ed. NY: Wiley, 2001. 2,083 p. (A Wiley-Interscience publication) 0471585890 QD251.2 547 00-32098

Revised ed. of *Advanced Organic Chemistry*, Jerry March, 4th ed., 1992. Contains some 600 reactions, structure information, stereochemistry, and general reaction mechanisms. The material is organized by reaction types, with chapters on aromatic nucleophilic and electrophilic substitution, free radical substitution, eliminations, rearrangements, oxidations, reductions, additions to carbon-carbon multiple bonds, and additions to carbon-hetero multiple bonds. Discusses scope, limitations and mechanisms, and provides extensive references to journal articles for detailed experimental procedures.

5.063

Sykes, Peter. *A guidebook to mechanism in organic chemistry*. 6th ed. Harlow, Essex, England; NY: Longman; NY: Wiley, 1986. 416 p. 0470206632; 0582446953 pbk QD476 547.1 85-24067

Classic work on mechanistic organic chemistry, with examples.

5.064

Vogel, Arthur Israel. *Vogel's elementary practical organic chemistry: preparations.* 3rd ed. rev. by B.V. Smith and M.M. Waldron. London; NY: Longman, 1980. 407 p. 0582470099 QD261 547 78-41266

(Part 1, small-scale preparations, 1966; part 2, qualitative organic analysis, 1st ed., 1957, 2nd ed., 1965; part 3: quantitative organic analysis, 1958; *Vogel's Elementary Practical Organic Chemistry: Preparations*, 3rd ed., rev. by B.V. Smith and M.M. Waldron, 1980.) Pt. 2 treats various practical aspects of organic chemistry, including determination of physical constants, reactions of organic compounds, qualitative analysis for elements, solubility classes, class reactions, preparation of derivatives, qualitative analysis of mixtures of organic compounds, physical constants of organic compounds, and tables of derivatives.

5.065

Vogel, Arthur Israel. *Vogel's textbook of practical organic chemistry*. 5th ed., rev. by B.S. Furniss [and others]. London: Longman Scientific & Technical; NY: Wiley, 1989. 1,514 p. 0470214147 QD261 547 88-36786

4th ed. published in 1978 as *Vogel's Textbook of Practical Organic Chemistry, Including Qualitative Organic Analysis*. 3rd ed., 1957/56. Treats general techniques including distillation, solutions of liquids in liquids, melting and freezing; experimental techniques and apparatus; preparation and reactions of various aliphatics, aromatics, heterocyclics, alicyclics, and miscellaneous reactions; organic reagents in organic and inorganic chemistry; dyes and indicators; physiologically active compounds; polymers; qualitative organic analysis; semimicro techniques; literature of organic chemistry.

5.066

Wiberg, Kenneth B. *Physical organic chemistry*. NY: Wiley, 1964. 591 p. QD476 547.1 63-20644

Contents: Bonding and spectra; Equilibria; Kinetics; appendixes with physical constants, spectral data, thermodynamic functions, and other materials.

STRUCTURE AND PROPERTIES: ELECTRONIC/PHYSICAL

5.067

Dewar, Michael J.S. *The molecular orbital theory of organic chemistry*. NY: McGraw-Hill, [1969]. 484 p. (McGraw-Hill series in advanced chemistry) QD461 547 68-21840

Discusses organic chemistry in terms of self-consistent field molecular orbital calculations rather than ones based on the Hückel molecular orbital method, and offers the perturbational molecular orbital treatment as an alternative to resonance theory.

5.068

Handbook of chemical property estimation methods for chemicals: environmental and health sciences, [ed. by] Robert S. Boethling and Donald Mackay. Boca Raton, FL: Lewis Publishers, 2000. 1566704561 QD271 547 99-58377

A complete restructuring and updating of *Handbook of*

Chemical Property Estimation Methods: Environmental Behavior of Organic Compounds by Warren J. Lyman, William F. Reehl, and David H. Rosenblatt. Addresses melting and boiling points, vapor pressure, Henry's Law constant partitioning properties, octanol-water partition coefficients, water solubility, sorption, bioconcentration in fish, and partitioning to aerosols; estimation methods for degradation process rates, half-lives for biodegradation, hydrolysis, atmospheric oxidation, photolysis in water, and redox reactions; membrane penetration and diffusion; properties of surface active chemicals; and reviews chemical property estimation methods and their role in fate assessment.

5.069

Jorgensen, William L., and Lionel Salem. *The organic chemist's book of orbitals*. NY: Academic Press, 1973. 305 p. 0123902509; 0123902568 pbk QD461 547 72-9990

Chap. 1: How molecular orbitals are built by delocalization: a unified approach based on bond orbitals and group orbitals; Chap. 2: Basic data concerning the orbital drawings in chap. 3; Chap. 3: Three-dimensional molecular orbitals; Chap. 4: Index of references.

5.070

Molecular modelling and bonding, ed. by Elaine Moore. Cambridge, UK: Royal Society of Chemistry, 2002. 152 p. 1 CD-ROM (The molecular world) 0854046755 QD480 541.22

Discusses molecular and quantum mechanics, atomic and molecular orbitals, computational chemistry, and bonding in solids.

5.071

Organic electrochemistry, ed. by Henning Lund and Ole Hammerich. 4th ed., rev. and expanded. NY: M. Dekker, 2001. 1,393 p. 0824704304 QD273 547 00-47593

Presents new developments in electrochemistry of C_{60} and related compounds; electroenzymatic synthesis; conducting polymers; electrochemical partial fluorination; and updated information on carbonyl compounds; anodic oxidation of oxygen-containing compounds; electrosynthesis of bioactive materials; and electrolytic reductive coupling.

5.072

Quinkert, Gerhard, Ernst Egert, and Christian

Griesinger. *Aspects of organic chemistry*, [transl. by Andrew Beard]. Basel: Verlag Helvetica Chimica Acta; Weinheim; NY: VCH, 1996- . [v. 1] Structure. 3906390152 QD251.2

The first in a series of planned volumes in which authors take up in detail topics that are extra in introductory organic chemistry courses. This volume discusses aspects of stereochemistry in sufficient depth that the reader should be able to understand large biological molecules such as RNA, DNA, and proteins. A variety of other topics are also treated, such as the nature of aromaticity, uses of nuclear magnetic resonance (NMR) spectroscopy in structure determination, and biochemistry.

5.073

Rauk, Arvi. *Orbital interaction theory of organic chemistry*. NY: Wiley, 1994. 307 p. (A Wiley-Interscience publication) 0471593893 QD461 547.1 93-27021

Examines and explains a rather wide variety of organic chemical reactions through a primarily qualitative application of molecular orbital interaction theory. Chapters present fundamental concepts of symmetry and orbital theory necessary to understanding material that follows; uses orbital interaction theory to study reactivity of functional groups such as alkenes, carbonyl compounds, and aromatic compounds; investigates reactive intermediates, nucleophilic substitution, proton abstraction, pericyclic reactions, and photochemistry. Numerous clear, appropriate, and useful structures. Extensive references; comprehensive index. Includes an interactive computer program for performing simple Hückel molecular orbital calculations.

5.074

Smith, William B. *Introduction to theoretical organic chemistry and molecular modeling*. NY: Wiley-VCH Publishers, 1996. 192 p. 1560819375 QD476 547.1 95-42883

A revision of the earlier *Molecular Orbital Methods in Organic Chemistry: HMO and PMO* (1974), extending discussion of quantum theory to the topic of molecular modeling. Chapters treat traditional topics of quantum organic chemistry, and computer programming and modeling. The book is organized thus: hydrogen atom and molecule; Hückel molecular orbital theory; the perturbation molecular orbital (PMO) method; aromaticity, antiaromaticity, and reactions; Hückel and frontier molecular orbital (FMO) theories applied to chemical reactivity; pericyclic reactions, orbital symmetry, and PMO and FMO

theories; improvements and extensions of the Hückel theory; molecular modeling—molecular mechanics, semiempirical methods, ab initio and density functions.

5.075

The third dimension, ed. by Lesley Smart and Michael Gagan. Cambridge: Royal Society of Chemistry, 2002. 240 p. 2 CD-ROMs (The molecular world) 0854046607 QD921

Explores atomic arrangements in molecules and different types of solids. Describes common crystal structures; metallic, ionic, molecular, and extended covalent crystals; crystal defects; chirality; isomerism; and the stereochemical consequences of the tetrahedral carbon atom.

SYNTHETIC METHODS

Annuals and Other Serial Publications

5.076

Annual reports in organic synthesis. NY: Academic Press, 1970- . Annual. ISSN 0066-409x QD262 547 71-167779

Issue for 1988 never published. An annual review of synthetically useful information that groups related reactions together. Fast scanning is facilitated by use of structures and reaction schemes. A disadvantage is that each annual volume must be scanned, and there is no cumulative index. References to journal articles are listed, along with selective experimental notes. An excellent source for helping chemists stay up-to-date and to quickly scan for relevant reactions as needed. The detailed table of contents serves in place of an index.

5.077

Liotta, Dennis, and Mark Volmer. *Organic syntheses reaction guide: incorporating collective volumes 1-7 and annual volumes 65-68*. NY: Wiley, 1991. 854 p. 047154261x QD262 21-17747

5.078

Organic syntheses; ed. in chief, Henry Gilman. NY: Wiley, v. 1- , 1921- . v. 1-55, 1921-1975. v. 70, 1992. ISSN 0078-6209 QD262 547.058 21-17747

"An annual publication of satisfactory methods for the preparation of organic chemicals." Editors: 1921- , R.

Adams, J.B. Conant, H.T. Clarke, O. Kamm. Collected and indexed every ten volumes into "Collective volumes." Coll. v.1 (v.1-9), H. Gilman, ed. 580 p.; coll. v.2 (v.10-19), A.H. Blatt, ed. 654 p. 1943; coll. v.3 (v.20-29), E.C. Horning, ed. 890 p. 1955; coll. v.4 (v.30-39), N. Rabjohn, ed. 1,036 p. 1963; coll. v.5 (v.40-49), H.E. Baumgarten, ed. 1,234 p. 1973; coll. v.6 (v.50-59), W.E. Noland, ed. 1,209 p. 1988; coll. v.7 (v.60-64), J.P. Freeman, ed. 602 pp. 1990; Cumulative indexes (v.1-49), R.L. Shriner and R.H. Shriner, eds. 432 pp. 1976. Online version is available gratis.

5.079
REACCS [computer-readable database]. San Leandro, CA: Molecular Design Ltd., 1990.

The complete database includes the following files: Theilheimer, Synthetic Methods of Organic Chemistry, 1940-80; Journal of Synthetic Methods (from which the annual Theilheimer volumes are currently derived), 1980 to date; Organic Syntheses, 1921 to date; current literature file (reaction information of interest to synthetic chemists, abstracted from currently published articles), 1986 to date; CHIRAS (asymmetric chemistry; chiral reactions), 1975 to date; METALYSIS (articles about transition metal catalysis), 1975 to date; and Comprehensive Heterocyclic Chemistry, the eight-volume set (Pergamon, 1984). The database currently includes approximately 220,000 reactions, with 10,000-12,000 added annually. Useful for locating functional group reactions. Structures of reactants and products (either a functional group partial structure or a specific molecule) may be searched, with reaction sites, type of reaction (such as reduction), and stereochemistry specified. Reaction diagrams and structures of all reactants and products in each record permit inspection for relevance. Other information provided in tabular form includes reaction conditions, physical properties, yield, and literature references.

5.080
Sugasawa, Shigehiko, and Seijirō Nakai. *Reaction index of organic syntheses.* [Rev. ed.] Tokyo: Hirokawa; NY: Wiley, 1967. 269 p. QD262 547 67-6627

Index to v. 1-45 of *Organic Syntheses*.

Compendia of Synthetic Methods

5.081
Anand, Nitya, Jasjit S. Bindra, and Subramania

Ranganathan. *Art in organic synthesis.* 2nd ed. NY: Wiley, 1988. 427 p. (A Wiley-Interscience publication) 0471887382 QD262 547 87-14762

A source for research chemists, with summaries of major achievements in natural product syntheses. Useful for identifying characteristic uses of reagents in synthesis. Provides synthesis information for approximately 100 substances that are representative of groups of compounds. Uses concise, easy-to-scan graphic representations of reaction pathways; journal references are listed if additional information is desired. Subject, author, and reaction type indexes.

5.082
Buehler, Calvin Adam, and Donald Emanual Pearson. *Survey of organic syntheses.* NY: Wiley-Interscience, 1970-76. 2 v., 1,166 p. 047111670x (v. 1) QD262 547 73-112590

Collects principal methods for synthesis of main types of organic compounds. Arranged by functional group. Discusses value, limitations, and theory of reactions listed, and offers preparative details for some materials. Author and subject indexes.

5.083
Carruthers, W. *Some modern methods of organic synthesis.* 3rd ed. Cambridge: Cambridge University Press, 1986. 526 p. (Cambridge texts in chemistry and biochemistry) 0521322340; 0521311179 pbk QD262 547 85-21270

Aimed at advanced undergraduates and beginning graduate students. Updates and complements House; covers many reactions not included in House, but in less detail. Provides the most important features of reactions. Chapters treat formation of carbon-carbon single bonds and double bonds, synthetic applications of organoboranes and organosilanes, oxidation and reduction, reactions at unactivated C-H bonds, and the Diels-Alder and related reactions. Information such as oxidation of alcohols or catalytic hydrogenation can be found readily in the rather detailed subject index.

5.084
Comprehensive organic functional group transformations; eds. in chief, Alan R. Katritzky, Otto Meth-Cohn, and Charles W. Rees. Oxford; NY: Pergamon, 1995. 7 v. 0080406041 (set) QD262 547 95-31088

V. 1: Synthesis: carbon with no attached heteroatoms; v. 2:

Synthesis: carbon with one heteroatom attached by a single bond; v. 3: Synthesis: carbon with one heteroatom attached by a multiple bond; v. 4: Synthesis: carbon with two heteroatoms, each attached by a single bond; v. 5: Synthesis: carbon with two attached heteroatoms with at least one carbon-to-heteroatom multiple link; v. 6: Synthesis: carbon with three or four attached heteroatoms; v. 7: Indexes.

5.085

Comprehensive organic synthesis: selectivity, strategy and efficiency in modern organic synthesis, ed. in chief, B.M. Trost; deputy ed. in chief I. Fleming. Oxford, England; NY: Pergamon Press, 1991. 9 v., 10,400 p. 0080359299 QD262 547.2 90-26621

V. 1: Additions to C-X pi-bonds, Pt. 1. Provides a detailed survey of reactions that entail the 1,2-addition of nonstabilized carbanion equivalents of carbonyl, imino, and thiocarbonyl functionality. V. 2: Additions to C-X pi-bonds, Pt. 2. Deals mainly with the addition reactions of delocalized carbanions (enolates) and their synthetic relatives (metalloenamines, enol ethers, allyl organometallics) with carbonyl compounds, imines, and iminium ions. Major emphasis is placed on C-C bond-forming reactions such as aldol and Mannich reactions, and acylation reactions. V. 3: Carbon-carbon sigma-bond formation. Covers carbon-to-carbon single bond-forming reactions, but only those that do not involve additions to C-X pi-bonds. V. 4: Additions to and substitutions at C-C pi-bonds. Focuses on additions and the resulting substitutions at carbon-carbon pi-bonds. V. 5: Combining C-C pi-bonds; covers important pi-bond-dependent transformations: thermal, photochemical, and metal-catalyzed cycloadditions of every major type. V. 6: Heteroatom manipulation. Treats functional groups containing heteroatoms important in organic synthesis. Describes introduction of these groups and their relevant transformations, and discusses various aspects of chemoselectivity, regioselectivity, and stereoselectivity. V. 7: Oxidation. Covers all methods of oxidation in organic synthesis. V. 8: Reduction of functional groups, organized into reduction of C=X, X=Y, C=C, and C=C bonds. V. 9: Cumulative author and subject indexes.

Provides a thorough and critical account of organic synthesis in a clear and systematic way, covering the formation of carbon-carbon bonds, the introduction of heteroatoms, and heteroatom interconversions. A major aspect of this work is "selectivity": organized in terms of chemo-, regio-, diastereo- and enantio-selectivity. The text is supported by computer-drawn structural formulas, tables and figures.

5.086

Corey, E.J., and Xue-min Cheng. *The logic of chemical synthesis*. NY: Wiley, 1989. 436 p. 0471509795 QD262 547.2 89-5335

Describes the synthon-based strategic analysis of synthetic design pioneered by Corey and coworkers. An advanced-level book that describes the principles of retrosynthetic analysis and transform- and structure-based strategies. Its visual synopses of actual syntheses illustrate these principles and form a useful catalog of multistep syntheses in the literature.

5.087

Harrison, Ian T., and Shuyen Harrison. *Compendium of organic synthetic methods*. NY: Wiley-Interscience, 1971- . v. 1-[7], 1971-[92]. 047135550x (v. 1) QD262 547 71-162800

This ongoing series provides rapid retrieval from the literature, listing material by reaction type rather than by author name or publication date. Each updated volume presents the latest synthetic methods for preparation of monofunctional and difunctional compounds. Each volume covers selective preparations published over several years. The uniquely direct index at the front of each volume (there is no cumulative index) provides a rapid method of locating pages for preparation of one functional group from another.

5.088

House, Herbert O. *Modern synthetic reactions*. 2nd ed. Menlo Park, CA: W.A. Benjamin, 1972. 856 p. (The Organic chemistry monograph series) 0805345019 QD262 547 78-173958

Written for advanced undergraduates and beginning graduate students. A rather detailed survey of many important synthetic methods commonly used. Examples of chapters: metal hydride reduction, oxidation, halogenation, alkylation, aldol condensation, acylation. However, some important reactions, such as the Diels-Alder reaction and Claisen rearrangement, are not discussed; instead, readers are referred to other sources. Information provided: reaction mechanism, stereochemistry, reaction conditions, examples of the reaction. For detailed experimental procedures, readers are referred to journal articles and other handbooks. Multiple references to the same reaction are usually provided. An extensive subject index provides quick access to information.

5.089

Ireland, Robert E. *Organic synthesis*. Englewood Cliffs, NJ: Prentice-Hall, [1969]. 147 p. (Prentice-Hall foundations of modern organic chemistry series) 0136408397 QD262 547 73-76870

Discusses planning the synthesis of an organic molecule, based on classical syntheses of natural products and of compounds with unusual structures.

5.090

Larock, Richard C. *Comprehensive organic transformations: a guide to functional group preparations.* 2nd ed. NY: Wiley-VCH, 1999. 0471190314 QD262 547 98-37314

Although the primary organization is by functional groups being synthesized, each section is subdivided by major reaction types, such as reduction, oxidation, alkylation, substitution, elimination, coupling, halogenation, isomerization, etc. In some ways, this is an updated version of *Synthetic Organic Chemistry*, by Wagner and Zook (1953). Focus is on major preparative reactions that are general in scope and on literature references in primary journals. It is not exhaustive and specialized reactions are often not included. Review articles are often listed at the beginning of sections.

5.091

Larock, Richard C. *Comprehensive organic transformations on CD-ROM [electronic resource]; a guide to functional group preparations.* NY: Wiley-VCH, 1997. 1 CD-ROM 047118649x QD262 547 97-4539

This electronic product, based on the book by Larock (1989), offers easy browsing showing chemical reactions, indexes organized by starting materials and products, and addition of common names for functional groups to the indexes. Searching is most efficiently done by query, using templates (including transformations from: to), products, starting materials, book headings, journals, reagents and catalysts, and personal search terms.

5.092

Organic synthesis highlights, by Johann Mulzer ... [et al.]. Weinheim; NY: VCH, 1991-1998. 3 v. V. 2 ed. by Herbert Waldmann; v. 3 ed. by Mulzer and Waldmann. 3527279555 (v. 1: Weinheim); 0895739186 (v. 1: NY); 03527292004 (v. 2); 3527293787 (v. 2) pbk QD262 547.2 90-13003

Offers up-to-date overviews of organic synthesis

problems, research areas, and strategies. Articles are written by experts.

5.093

Preparation of alkenes: a practical approach, ed. by Jonathan M.J. Williams. NY: Oxford University Press, 1996. 253 p. (The practical approach in chemistry series) 0198557957; 0198557949 pbk QD305.H7 547 96-13456

Alkenes are a common class of organic compounds for which a large number of preparative methods are known. Accessible chapters include detailed stepwise instructions for every procedure discussed, along with full lists of equipment, chemicals needed, and safety precautions for each. Other useful features (especially for novices) include an excellent introductory chapter describing general laboratory techniques, a list of chemical suppliers in the US and abroad, and a list of abbreviations. Summary chapter; chart to find procedures.

5.094

Smit, W.A., A.F. Bochkov, and R. Caple. *Organic synthesis: the science behind the art.* Cambridge, UK: Royal Society of Chemistry, 1998. 477 p. 0854045449 QD262

Presents the general idea of synthesis from the beginning of the 20th century to the present. Details outstanding achievements of modern organic synthesis, for scientific merit, aesthetic appeal of target molecules, and intrinsic beauty of the solutions to the problems.

Special Techniques in Synthesis

5.095

Asymmetric synthesis, ed. by Robert Alan Aitken and S. Nicholas Kilényi. London; NY: Blackie Academic & Professional, 1992. 233 p. 0412024519 QD262 547 91-46601

Explains general methods and terminology used, and describes the most important methods presently available for the formation of chiral compounds. Case studies show basic methods used in preparing some important naturally occurring and biologically active chiral compounds. Contents: Chirality: the description of stereochemistry; Analytical methods: determination of enantiomeric purity; Sources and strategies for the formation of chiral compounds; First- and second generation methods: chiral starting materials and auxiliaries; Asymmetric total synthesis: asymmetric reagents and catalysts.

5.096

Asymmetric synthesis, ed. by James D. Morrison. NY: Academic Press, 1983-85. 5 v. 0125077017 (v. 1) QD481 541.3 83-4620

V. 1: Analytical methods; v. 2-3: Stereodifferentiating addition reactions; v. 4: The chiral carbon pool and chiral sulfur, nitrogen, phosphorus, and silicon centers, ed. by James D. Morrison and John W. Scott; v.5: Chiral catalysis.

5.097

Giese, Bernd. *Radicals in organic synthesis: formation of carbon-carbon bonds.* Oxford; NY: Pergamon, 1986. 294 p. (Organic chemistry series, v. 5) 0080324932; 0080324940 pbk QD262 547 86-17052

Basic principles: A. General aspects of synthesis with radicals; B. Elementary reaction steps between radicals and non-radicals; C. Comparison of radicals and ions in synthesis. Intermolecular formation of aliphatic C-C bonds; Intermolecular formation of aliphatic C-C bonds; C-C bond formation of aromatic systems; Methods of radical formation.

5.098

Greene, Theodora W., and Peter G.M. Wuts. *Protective groups in organic synthesis.* 3rd ed. NY: Wiley, 1999. 779 p. (A Wiley-Interscience publication) 0471160199 QD262 547.2 98-38182

A desk or lab reference with detailed indexes and tables of contents, organization attuned to chemists' questions, and extensive graphics. The first chapter discusses basic aspects of protective group chemistry: properties of a protective group, development of new protective groups, how to select a protective group from those described in this book, and an illustrative example of the use of protective groups. Other chapters are organized by functional group to be protected: hydroxyl, phenol, carbonyl, thiol, and amino. A detailed table of contents at the beginning of each chapter helps identify relevant sections. Each chapter discusses chemistry of the available protective groups. Graphic structures, brief reaction schemes, reaction conditions, and yields offer fast visual scanning. The best methods of formation and cleavage are described and thoroughly referenced to journal articles. For each functional group to be protected there is a reaction chart that tries to predict reactivities (i.e., stabilities) of the main protective groups against several acids, bases, oxidants,

and other reagents and reaction conditions. Includes an extensive index.

5.099

Hassner, Alfred, and C. Stumer. *Organic syntheses based on name reactions.* 2nd ed. Amsterdam; Boston: Pergamon, 2002. 443 p. (Tetrahedron organic chemistry series, v. 22) 0080432603; 008043259x pbk QD262 547 2002-24278

(Revision of *Organic Syntheses Based on Name Reactions and Unnamed Reactions*, 1994.) Focuses on ongoing development in this area, and reflects the idea that many new reagents and reactions are now being referred to by their names. Includes some 540 stereoselective and regioselective reagents or reactions including asymmetric syntheses. Name, reagent, reaction, and functional transformation indexes.

5.100

Kyriacou, Demetrios K. *Basics of electroorganic synthesis.* NY: Wiley, 1981. 153 p. (A Wiley Interscience publication) 0471079758 QD273 547 80-25326

Discusses the fundamentals of organic synthesis using electrochemistry. Describes general electroorganic synthesis and techniques involved, briefly surveys electroorganic reactions, and covers some special topics (amalgams, organometallics, etc.), and concludes with probes of electroorganic synthesis.

5.101

Nógrádi, Mihály. *Stereoselective synthesis.* 2nd, thoroughly rev. and updated ed. Weinheim; NY: VCH, 1995. 368 p. 352729242x; 3527292438; 1560818956 QD481 541.2 97-165615

Methods and reagents for stereoselective synthesis are featured. New edition contains material on homogeneous diastereoselective hydrogenations, enantioselective oxidations, and novel, efficient chiral auxiliaries. Figures and tables.

5.102

Noyori, Ryoji. *Asymmetric catalysis in organic synthesis.* NY: Wiley, 1994. 378 p. (The George Fisher Baker non-resident lectureship in chemistry at Cornell University) (A Wiley-Interscience publication) 0471572675 QD262 547 93-3884

Discusses major methods of asymmetric catalysis with

numerous examples from different areas of organic chemistry. Includes homogeneous and heterogeneous asymmetric catalysis, chiral metal complexes, isomerization of olefins, chirality transfer, and other topics.

5.103

Ono, Noboru. *The nitro group in organic synthesis*. NY: Wiley-VCH, 2001. 372 p. (Organic nitro chemistry series) 0471316113 QD262 547 00-47762

The nitro group was one of the first groups in organic chemistry to be studied and investigated in the late 19th century. A few, but not many, reactions of nitro groups were known at that time. Consequently, most organic chemists tend to regard compounds as relatively uninteresting from a synthetic point of view, focusing instead on their electronic properties. The author shows that many new synthetic procedures have been developed that rely on nitro compounds as key intermediates.

5.104

Procter, Garry. *Asymmetric synthesis*. Oxford; NY: Oxford University Press, 1996. 237 p. 0198557264; 0198557256 pbk QD262 547 95-35780

Asymmetric synthesis is a technique whereby an additional substance is added to the reaction medium to divert the reaction into the desired course. Condenses the chief concepts; clearly explains the principles. Offers examples from recent literature.

5.105

Sandler, Stanley R., and Wolf Karo. *Organic functional group preparations*. 2nd ed. NY: Academic Press, 1983-89. 3 v. (Organic chemistry, v. 12) 0126186014 (v. 1) QD262 547 83-2555

Briefly describes typical experimental procedures for preparation of specific functional groups. Covers typical, classical reactions; not comprehensive. A given functional group will have 10-20 methods exemplified and therefore serves as a good introduction to its preparation. Each chapter covers a functional group, with information divided by type of reaction. The three volumes cover more than 50 functional groups. Many preparations are for monofunctional substances; little effort is made to ensure chemoselectivity. Lists selected additional preparative methods not covered in the text, along with journal references. Each volume includes an index for its contents; there is no overall index for all three volumes.

5.106

Santelli, Maurice, and Jean-Marc Pons. *Lewis acids and selectivity in organic synthesis*. Boca Raton, FL: CRC Press, 1996. 334 p. (New directions in organic and biological chemistry) 0848378664 QD262 547.2 95-35887

Lewis acids are an important class of compounds, primarily because of their growing use as catalysts for a wide variety of chemical reactions. A credible survey of the field of Lewis acid chemistry, with special attention to stereoselectivity in an assortment of organic reactions.

5.107

Seneci, Pierfausto. *Solid phase synthesis and combinatorial technologies*. NY: John Wiley & Sons, 2000. 637 p. 0471331953 QD262 547 99-86954

Reviews solid-phase synthesis, combinatorial chemistry, and related technologies in the areas of pharmaceuticals, biotechnology, materials science, catalysis, and agrochemistry. Available as an online work.

5.108

Stowell, John Charles. *Carbanions in organic synthesis*. NY: Wiley, 1979. 247 p. (A Wiley-Interscience publication) 047102953x QD305.C3 547 79-373

Discusses methods of preparing carbanions, their general reaction types, and specific reactions of carbanions of increasing delocalization.

5.109

Tsuji, Jiro. *Transition metal reagents and catalysts: innovations in organic synthesis*. Chichester; NY: Wiley, 2000. 477 p. 0471634980 QD505 547 2001-271784

Includes industrial processes using homogeneous transition metal catalysts; basic chemistry of transition metal complexes; reactions of organic halides and pseudohalides; reactions of allylic compounds; reactions of conjugated dienes; reactions of propargylic compounds; reactions of alkenes and alkynes; synthetic reactions via transition metal carbene complexes; protection and activation by coordination; catalytic hydrogenation, transfer hydrogenation, and hydrosilylation; reactions promoted and catalyzed by Pd(II) compounds.

Organic Chemistry

REACTIONS

5.110

CASREACT [computer-readable database]. *User guide*. Chemical Abstracts Service, 1989.

A chemical reactions database derived from more than 100 journals covered by Chemical Abstracts (CA) beginning with 1985. Only articles covered in the organic sections (Sections 21-34) of CA are included; currently there are more than 1 million reactions, with approximately 100,000 reactions added each year. The extensive database permits comprehensive searches. CASREACT is very useful for locating functional group reactions. Functional group partial structures and specific molecules can be built in either Registry or CASREACT files. A functional group partial structure or a specific molecule can be searched as a reactant or product. Reaction diagrams and structures for all reactants and products enable visual inspection for relevance. Other information provided includes reaction conditions, safety and hazard information, yields, registry numbers for all substances in the reactions, and literature reference. CASREACT is searchable on the STN International search service.

5.111

Grossman, Robert B. *The art of writing reasonable organic reaction mechanisms*. 2nd ed. NY: Springer, 2003. 355 p. 0387954686 QD502.5 547 2002-24189

Demonstrates the method of producing a reasonable mechanism for an organic chemical transformation. Organized by types of mechanisms and conditions under which the reaction takes place, rather than by the overall reaction. Suggests practical tips for drawing common mechanistic pathways.

5.112

Haines, Alan H. *Methods for the oxidation of organic compounds: alkanes, alkenes, alkynes, and arenes*. London; Orlando: Academic Press, 1985. 388 p. (Best synthetic methods) 0123155010 QD281.09 547 84-12465

5.113

Kharasch, Morris Selig, and Otto Reinmuth. *Grignard reactions of nonmetallic substances*. NY: Prentice-Hall, 1954. 1,384 p. (Prentice-Hall chemistry series) QD77 547.013 54-7458

5.114

Laue, Thomas, and Andreas Plagens. *Namen- und*

Schlagwort-Reaktionen der organischen Chemie = Named organic reactions. Stuttgart: Teubner, 1994; Chichester; NY: Wiley, 1998. 288 p. 0471971421 QD291 547.2 94-225755 (German edition)

Reprinted with revisions in March 2000. Offers 134 named reactions, well described, with up-to-date references.

5.115

Li, Chao-Jun, and Tak-Hang Chan. *Organic reactions in aqueous media*. NY: Wiley, 1997. 199 p. 0471163953 QD255.4 547 96-29886

Up-to-date treatment of organic reactions taking place in water. Hydrolysis reactions are omitted. Treats fundamental properties of water; pericyclic reactions, including the Diels-Alder reaction; nucleophilic additions and substitutions; metal-mediated reactions; transition-metal catalyzed reactions; oxidations and reductions; and industrial applications.

5.116

Li, Jie Jack. *Name reactions: a collection of detailed reaction mechanisms*. 2nd ed., [rev. and enl.] Berlin; NY: Springer, 2003. 465 p. 3540402039 QD291 547.2

More than 300 reactions are listed together with stepwise mechanisms; references include the original along with review articles through 2001. Contemporary reactions, such as asymmetric syntheses, are included.

5.117

Moody, Christopher J., and Gordon H. Whitham. *Reactive intermediates*. Oxford; NY: Oxford University Press, 1992. 89 p. (Oxford chemistry primers; 8) 019855673x; 0198556721 pbk QD476 547.2 92-12267

Describes uncharged reactive species—radicals, carbenes, nitrenes, arynes (benzyne)—and their synthetic utility.

5.118

Miller, Audrey, and Philippa H. Solomon. *Writing reaction mechanisms in organic chemistry*. 2nd ed. San Diego: Harcourt/Academic Press, 2000. 471 p. (Advanced organic chemistry series) 0124967124 QD502.5 547 99-61410

Discusses mechanistic pathways from fundamental chemical, physical, and electronic/molecular orbital principles. Mechanisms discussed include reactions of nucleophiles and bases, electrophiles and acids, radicals and radical ions, and pericyclic reactions.

5.119

Mundy, Bradford P., and Michael G. Ellerd. *Name reactions and reagents in organic synthesis.* NY: Wiley, 1988. 546 p. (A Wiley-Interscience publication) 0471836265 QD291 547 88-14915

Compiles name reactions and reagents; reactions are listed alphabetically, with (either) adopted or proposed mechanism, secondary information, and some referenced examples. A second section lists reagents alphabetically, with structure, physical properties, major uses, preparation and commercial availability, precautions, secondary information, and some referenced examples. Reagents are cross-referenced to Fieser & Fieser, *Reagents for Organic Synthesis*, through v. 12.

5.120

Olah, George A. *Friedel-Crafts and related reactions.* NY: Interscience Publishers, 1963-65. 4 v. in 6. QD501 547.2082 63-18351

V. 1: General aspects; v. 2: Alkylation and related reactions (2v.); v. 3: Acylation and related reactions (2v.); v. 4: Miscellaneous reactions.

5.121

Organic reactions. Original editor, Roger Adams. NY: John Wiley & Sons. v. 1- , 1942- . Irregular. ISSN 0078-6179 QD251 547 42-20265

"Critical discussion of widely used organic reactions or partial phases of a reaction." Editorial board [v. 1-] Roger Adams, ed. in chief, Werner E. Buchmann, Louis F. Fieser [and others]. V. 50 published in 1997.

5.122

Patai, Saul. *Patai's guide to the chemistry of functional groups.* Chichester, England; NY: Wiley, 1989. 455 p. (An Interscience publication) 0471915262 QD251.2 547 88-20707

This volume serves as an index to volumes from 1946 to 1988 of *Chemistry of Functional Groups*, ed. by S. Patai.

5.123

Pross, Addy. *Theoretical and physical principles of organic reactivity.* NY: Wiley, 1995. 294 p. (A Wiley-Interscience publication) 0471555991 QD476 547.1 95-21389

Discusses theoretical principles, principles of physical organic chemistry, and reaction types. The author presents a more unified approach to organic reactivity that merges qualitative theoretical approaches with traditional concepts of physical organic reactivity and then applies them to the major types of organic reaction mechanisms.

5.124

Reactivity and structure: concepts in organic chemistry. Berlin; NY: Springer-Verlag, v. 1- , 1975- . ISSN 0341-2377 82-3953

Series critically evaluates, compiles, and integrates the important information on contemporary research issues. Each monograph deals with the problem of the reactivity and structure of organic compounds, including synthetic methods and theory.

ORGANIC CHEMISTRY OF SPECIAL ELEMENTS

5.125

Brook, Michael A. *Silicon in organic, organometallic, and polymer chemistry.* NY: Wiley, 2000. 680 p. (A Wiley-Interscience publication) 0471196584 QD412.S6 547 98-49722

This comprehensive review of organosilicon chemistry updates C. Eaborn's original monograph *Organosilicon Compounds* (1960). Introduction and extensive citation lists to important topics in silicon chemistry. Topical organization emphasizes preparative organic/organosilicon chemistry, including silicones and other technologically important silicon polymers. Expands traditional organosilicon chemistry to include silicon-organometallics and silicon in biological systems.

5.126

Brown, Herbert C. *Hydroboration.* NY: W.A. Benjamin, 1962. 290 p. QD281 547.23 62-12322

Brown received the Nobel prize for his work on this reaction. Book contents: Introduction and survey; Early history; Chemistry of organoboranes; Borohydride chemistry; Hydroboration procedures; Scope; Directive effects; Stereochemistry; Isomerization; Displacement reactions; Hydroboration of hindered olefins; Alkylboranes; Selective hydroboration with disiamyl (disecondary isoamyl)borane; Asymmetrical hydroboration with diisopinocampheyl borane; Hydroboration of dienes; Hydroboration of acetylenes; Diborane as a reducing agent;

Disiamylborane as a reducing agent; Hydroboration of functional derivatives.

5.127

Chemistry of organic fluorine compounds II: a critical review, ed. by Milos Hudlicky and Attila E. Pavlath. Washington, DC: American Chemical Society, 1995. 1,296 p. (ACS monograph, 187) 084122515x QD305.H15 547 95-20195

Not a replacement for Hudlicky's *Chemistry of Organic Fluorine Compounds: A Laboratory Manual with Extensive Literature Coverage* (2nd rev. ed., 1992), this is a supplement with new material covering literature from 1972 to 1991. The organization parallels that of the earlier work, with chapters (written by the editors and 44 contributors) on fluorinating agents, synthesis of organic fluorine compounds, their reactions, uses as reagents, properties, analysis including NMR, and applications. Besides nearly 4,500 literature citations, there is a useful compilation of monographs and review articles in the first chapter. Many illustrations using well-drawn structures; 50-page subject index; 100-page author index.

5.128

Colvin, Ernest W. *Silicon reagents in organic synthesis.* San Diego, CA: Academic Press, 1988. 147 p. (Best synthetic methods) 0121825604 QD412 547.2 89-164624

Twenty chapters. Introduction; review articles and suppliers; vinylsilanes; alpha,beta-epoxysilanes; allyl silanes; aryl silanes; alkynyl and propargyl silanes; silyl anions; oxidative cleavage via organofluorosilicates; Peterson olefination; beta-ketosilanes; acylsilanes; aminosilanes; alkyl silyl ethers; silol enol ethers and ketene acetals; silyl-based reagents; silanes as reducing agents; organic syntheses; organic reactions; organometallic syntheses; index of compounds and methods.

5.129

Goldwhite, Harold. *Introduction to phosphorus chemistry.* Cambridge; NY: Cambridge University Press, 1981. 113 p. (Cambridge texts in chemistry and biochemistry) 0521229782; 0521297575 pbk QD181.P1 546 79-27141

Good, brief introduction to the organic chemistry of phosphorus. Little mention of inorganic phosphorus chemistry.

5.130

Hudlicky, Milos. *Chemistry of organic fluorine compounds: a laboratory manual with comprehensive literature coverage.* 2nd (rev. ed.). NY: Ellis Horwood; PTR Prentice Hall, 1992. 903 p. (Ellis Horwood PTR Prentice Hall organic chemistry series) 0131316737 QD305 547 92-241426

Bibliography, pp. 729-859. 1st ed., 1962. Introduction; Apparatus; Reagents; Methods to introduce fluorine into compounds; Preparation of fluoroorganic compounds; Reactions of organic fluorine compounds; Fluorine compounds as chemical reagents; Properties of organic fluorine compounds; Analysis of organic fluorides; Practical applications.

5.131

Quin, Louis D. *A guide to organophosphorus chemistry;* structural illustrations by Gyöngyi Szakál Quin. NY: Wiley, 2000. 394 p. 0471318248 QD305.P46 547 99-43429

Surveys most important phosphorus-containing functional groups; syntheses; stereochemistry; coordination states, heterocycles, isotopic applications, electronic mechanisms, reactive intermediates, and spectroscopy. Discusses phosphorus compounds in living systems and agriculture.

5.132

Thomas, Susan E. *Organic synthesis: the roles of boron and silicon.* Oxford; NY: Oxford University Press, 1991. 91 p. (Oxford chemistry primers; 1) 0198556632; 0198556624 pbk QD412.B1 547 91-13680

Discusses the important uses of organoboron and organosilicon compounds for resolution of problems in organic synthesis, including their use in synthesis of homochiral molecules. Discussions of reactions and mechanisms are accompanied by illustrated equations. Reprinted with corrections, 1993.

SPECIAL TOPICS IN ORGANIC CHEMISTRY

Spectroscopy

5.133

Atlas of spectral data and physical constants for organic compounds. Grasselli, Jeanette G. and William M. Ritchey, editors. 2nd ed. Cleveland, OH: CRC Press, 1975. 6 v. 0878193170 QD257.7 547 75-2452

1st ed., 1973, edited by Grasselli, has title: *CRC Atlas of Spectral Data and Physical Constants for Organic Compounds*. Tabular compilation of spectra from other published sources. V. 1 contains names and synonyms, structures, and spectroscopic aids; v. 2-4 contain data tables; v. 5-6 contain indexes of molecular weight, physical constants, WLNs, mass spectra, and other constants.

5.134

Cooper, James William. *Spectroscopic techniques for organic chemists*. NY: Wiley, 1980. 376 p. (A Wiley-Interscience publication) 0471051667 QD272.S6 547 79-23952

Introduces spectroscopic techniques most frequently used in structure elucidation. Discusses infrared, NMR, Fourier transform NMR, 13C NMR, ultraviolet, Raman, and mass spectroscopy.

5.135

Dollish, Francis R., William G. Fateley, and Freeman F. Bentley. *Characteristic Raman frequencies of organic compounds*. NY: Wiley, 1974. 443 p. (A Wiley-Interscience publication) 0471217697 QC462.85 547 73-12157

Includes 108 representative Raman spectra.

5.136

Field, L.D., S. Sternhell, and J.R. Kalman. *Organic structures from spectra*. 2nd ed. Chichester; NY: John Wiley, 1995. 205 p. 0471956309; 0471956317 pbk QD272 547.3 95-31574

Rev. ed. of *Organic Structures from Spectra* by S. Sternhell and J.R. Kalman, 1986. Adds some 70 new problems, most featuring NMR spectra obtained at higher fields than in the 1st ed., DEPT experiments, and coupled 13C NMR spectra.

5.137

Friedel, Robert A., and Milton Orchin. *Ultraviolet spectra of aromatic compounds*. NY: Wiley, 1951. 52 p. QC459 535.84 51-8747

The 600 spectra are indexed by name and formula.

5.138

The infrared spectra handbook of common organic solvents. Philadelphia, PA: Sadtler, 1983. 400, 400, 28, 27 p. 0845600931 QC463 547.1 83-50550

Spectra of 400 most commonly used solvents, in a transmittance vs. wavenumber format. Data include physical constants, CASRN, NIOSH number, and information on use, solubility, flammability, and toxicity.

5.139

The infrared spectra handbook of intermediates. Philadelphia, PA: Sadtler, 1987. 15, 491, 42, 11, 11 p. 0845601334 QC462 547.3 86-61363

Contains nearly 500 IR absorption spectra arranged in 17 major classes of compounds. The classes included are carboxylic acids, their anhydrides, acid halides, esters, and amides; sulfonic acid derivatives; phosphorus-containing compounds; isocyanic and isothiocyanic acid esters; and aldehydes, nitriles, ketones, alcohols, thiols, amines, ethers and thioethers, halogenated hydrocarbons, and hydrocarbons.

5.140

Klessinger, Martin and Josef Michl. *Excited states and photochemistry of organic molecules*. NY: VCH, 1995. 537 p. 1560815884 QD476 547.1 92-46464

In this work the approach is nonmathematical and qualitative; it provides a qualitative introduction to electronic excitation of organic molecules and to electronic spectroscopy, photochemistry, and photophysics. Emphasizes potential energy surfaces, bonding theory, and molecular electronic structures and recent developments in roles of conical intersections in understanding photochemical and photophysical processes. Chapters present basic electronic spectroscopy for nonspecialists; discuss absorption spectra of the most important organic molecules; examine natural circular dichroism and magnetic circular dichroism; introduce concepts of potential energy surfaces, barriers, minima, and funnels; discuss physical and chemical transformations of excited states: photophysical processes, photochemical reactions in condensed media, and phototransformations of organic molecules.

5.141

Organic electronic spectral data. NY: Wiley & Sons. v. 1- , 1946/52- . (An Interscience publication) ISSN 0078-6136 QC437 547.346 60-16428

Editors: 1946/52, M.J. Kamlet; 1953/55, H.E. Ungnade. Arranged by molecular formula. Gives name, solvent or phase, and wavelength values. References to literature.

5.142

Organic structural spectroscopy, by Joseph B. Lambert et al. Upper Saddle River, NJ: Prentice Hall, 1998. 568 p. 0132586908 QD272.S6 547 97-40522

Discusses and explains the most common and useful spectroscopic techniques for precise structure determination. Covers NMR, IR, UV-VIS, Raman, mass, and chiroptical spectroscopies.

5.143

Perkampus, Heinz-Helmut. *UV-VIS-Spektroskopie und ihre Anwendungen*. Berlin; NY: Springer Verlag, 1986. 208 p. (Anleitungen für die chemische Laboratoriumspraxis, Bd. 21) 3540154671 (Berlin); 0387154671 (NY) QD96

An essential tool for all working with UV-VIS spectroscopy, this atlas provides a systematic and critical selection of reproduced spectra important to analytical problems (e.g., in HPLC with UV detection) in chemistry, physics, biology, and clinical chemistry.

5.144

Silverstein, Robert M., and Francis X. Webster. *Spectrometric identification of organic compounds*. 6th ed. NY: Wiley, 1998. 482 p. 0471134570 QD272 547 97-21336

1st ed. published in 1962. Identification using mass, infrared, and NMR spectrometry.

Stereochemistry

5.145

Buckingham, J., and R. A. Hill. *Atlas of stereochemistry: absolute configurations of organic molecules. Supplement.* 2nd ed. NY: Chapman and Hall, 1986. 309 p. 041226000x QD481 541.2 85-12825

Supplement to: *Atlas of Stereochemistry: Absolute Configurations of Organic Molecules* by W. Klyne and J. Buckingham, 2nd ed., London, 1978. Contents: Fundamental chiral compounds; Carbohydrates; Terpenes (including steroids); Alkaloids; Miscellaneous natural products; Compounds with chirality due to isotopic substitution; Compounds containing chiral axes, planes, etc.; Compounds containing chiral atoms other than carbon; Cumulative author, subject, molecular formula indexes.

5.146

Conformational analysis, by Ernest L. Eliel ... [et al.].

Washington, DC: American Chemical Society, 1981, c1965. 524 p. Reprint. Originally published: NY: Interscience Publishers, 1965. 0841206538 QD481 547.1 81-10783

A valuable tool for organic research that treats conformations of organic molecules, as distinguished from their structures or configurations. Treats cyclohexanes, other ring systems, natural products (steroids, triterpenoids, alkaloids), and carbohydrates.

5.147

Eliel, Ernest L., Samuel H. Wilen, and Michael P. Doyle. *Basic organic stereochemistry*. NY: Wiley-Interscience, 2001. 688 p. 0471374997 pbk QD481 547 2001-17847

Based on *Stereochemistry of Organic Compounds*, 1994 (see entry following). Introduction; Structure; Stereoisomers; Symmetry; Configuration; Properties of stereoisomers and stereoisomer discrimination; Separation of stereoisomers, resolution, and racemization; Heterotropic ligands and faces: prostereoisomerism and prochirality; Stereochemistry of alkenes; Conformation of acyclic molecules; Configuration and conformation of cyclic molecules; Chiroptical properties; Chirality in molecules devoid of chiral centers.

5.148

Eliel, Ernest L., and Samuel H. Wilen. *Stereochemistry of organic compounds*, with a chapter on stereoselective synthesis by Lewis N. Mander. NY: Wiley, 1994. 1,267 p. (A Wiley-Interscience publication) 0471016705 QD481 547.1 93-12476

Completely revised and updated version of Eliel's classic work, *Stereochemistry of Carbon Compounds*, 1962. Extensive coverage of conformational analysis in cyclic and acyclic systems. Discusses asymmetric and diastereoselective synthesis, conformational analysis, properties of enantiomers and racemates, separation and analysis of enantiomers and diastereoisomers, developments in spectroscopy, prostereoisomerism, concepts of stereochemistry, and chiroptical properties. Includes more than 1,000 illustrations and chemical structures, tables, more than 2,500 references, and a glossary.

5.149

Flapan, Erica. *When topology meets chemistry: a topological look at molecular chirality*. Cambridge; NY: Cambridge University Press; Washington, DC:

Mathematical Association of America, 2000. 241 p. (Outlooks) 0521662540; 0521664829 pbk QD455.3.T65 541 00-27517

Treats basic knot theory, three-dimensional manifolds, and the topology of embedded graphs in space. Discusses how topology links the understanding of molecular structures and how it might apply to chemistry and molecular biology.

5.150

Kagan, Henri. *Organic stereochemistry*, tr. by M.C. Whiting and U.H. Whiting. NY: Wiley, [1979]. 166 p. (A Halsted Press book) 0470267259 QD481 547 79-11500

Defines important terms as introduced, illustrates them with drawings, and offers examples. Point groups and symmetry are concisely and well covered, and applied to chirality and molecular symmetry. Discusses the concept of prochirality.

5.151

Klyne, William and J. Buckingham. *Atlas of stereochemistry: absolute configurations of organic molecules*. 2nd ed. London: Chapman and Hall, 1978. 2 v. 0195200586 (v. 1) QD481 547 78-2366

Supplemented by Buckingham, J. and R. A. Hill, *Atlas of stereochemistry: Absolute Configurations of Organic Molecules. Supplement*, 1986.

5.152

Mezey, Paul G. *Shape in chemistry: an introduction to molecular shape and topology*. NY: VCH, 1993. 224 p. 0895737272 (NY); 3527279326 (Weinheim) QD461 541.2 93-15622

A moderately rigorous treatment of chemical shape. Chapters treat intuitive concept of molecular shape; quantum chemical concept of molecular shape; applied topology; molecular bodies and surfaces and their topological representations; topological shape groups, codes, graphs, matrices, and globes; quantification of molecular similarity and complementarity; quantitative shape-activity relations in drug design and engineering; and special topics relating to symmetry, syntopy, and symmorphy.

5.153

Mislow, Kurt. *Introduction to stereochemistry*. NY: W.A. Benjamin, 1965. 193 p. (The Organic chemistry monograph series) QD481 547.1223 65-10940

Second printing with corrections, 1966. In three major sections: Structure and Symmetry, Stereoisomerism, and Separation and Configuration of Stereoisomers.

5.154

Morris, David G. *Stereochemistry*. NY: Wiley-Interscience; [Cambridge, England]: Royal Society of Chemistry, 2002. 170 p. (Basic concepts in chemistry) 0471224774 pbk QD481 547 2002-511021

Introduces special topics of use in stereochemistry. Treats diastereoisomers, bicyclic compounds, stereochemistry of imines, chiral compounds lacking a stereogenic center, heteroatoms as centers of chirality, prochirality, and other special topics and terminology.

5.155

Ramsay, O. Bertrand. *Stereochemistry*. London; Philadelphia: Heyden, 1981. 256 p. (Nobel prize topics in chemistry) 0855016817; 0855016825 pbk QD481 541.2 82-106278

One in a series of historical monographs related to the Nobel Prize in chemistry. Presents key works that led to the Prize for various chemists working in stereochemistry.

5.156

Topics in stereochemistry, ed. by Norman L. Allinger and Ernest L. Eliel. NY: John Wiley & Sons, 1967- . ISSN 0082-500x QD481 547 67-13943

Indexes to v. 1-4, 1967-69 in v. 4; v. 1-5, in v. 5. This series is designed to bridge the gap between works in stereochemistry and the current literature and to treat in greater detail some topics discussed in standard works on the subject.

Heterocyclics

5.157

Comprehensive heterocyclic chemistry: the structure, reactions, synthesis and uses of heterocyclic compounds. Ed. board: Alan R. Katritzky, chairman; Charles W. Rees, co-chairman. Oxford & NY: Pergamon, 1984. 8 v. 0080262007 (set) QD400 547.59 83-4264

Contents: v. 1: Introduction, nomenclature, review literature, biological aspects, industrial uses, less-common heteroatoms. V. 2: Six-membered rings with one nitrogen atom. V. 3: Six-membered rings with oxygen, sulfur, or two or more nitrogen atoms. V. 4: Five-membered rings with one oxygen, sulfur, or nitrogen atoms. V. 5: Five-membered rings with two or more nitrogen atoms. V. 6: Five membered rings with two or more oxygen, sulfur, or nitrogen atoms. V. 7: Small and large rings. V. 8: Author, subject, ring, and data indexes.

5.158

Comprehensive heterocyclic chemistry II: a review of the literature 1982-1995: the structure, reactions, synthesis, and uses of heterocyclic compounds. Eds. in chief Alan R. Katritzky, Charles W. Rees, Eric F.V. Scriven. Oxford; New York: Pergamon, 1996. 11 v. in 12 0080420729 (set) QD400 547.59 96-17870

Includes bibliographic references and index. V. 1A: Three-membered rings, with all fused systems containing three-membered rings; v. 1B: Four-membered rings, with all fused systems containing four-membered rings; v. 2: Five-membered rings with one heteroatom and fused carbocyclic derivatives; v. 3: Five-membered rings with two heteroatoms and fused carbocyclic derivatives; v. 4: Five-membered rings with more than two heteroatoms and fused carbocyclic derivatives; v. 6: Six-membered rings with two or more heteroatoms and fused carbocyclic derivatives; v. 7: Fused five- and six-membered rings without ring junction heteroatoms; v. 8: Fused five- and six-membered rings with ring junction heteroatoms; v. 9: Seven-membered and larger rings and fused derivatives; v. 10: Author and ring indexes; v. 11: Subject index.

5.159

Davies, David T. *Aromatic heterocyclic chemistry.* Oxford; NY: Oxford University Press, 1991. 88 p. 0198556616; 0198556608 pbk QD400 547 91-34831

Discusses the essential features of nine of the most important heterocyclic ring structures. Emphasizes mechanisms of reactions and explains retrosynthetic analysis used to understand the synthesis of heterocycles.

5.160

Katritzky, Alan R., in collaboration with C.W. Bird [et al.]. *Handbook of heterocyclic chemistry.* Oxford; NY: Pergamon, 1985. 542 p. 0080262171; 0080307264 pbk QD400 547 84-25492

Provides a one-volume overall picture of the largest division of organic chemistry. Contents: Preliminaries.

Structure of heterocycles: overview; six-membered rings; five-membered rings with one heteroatom; five-membered heterocycles with two or more heteroatoms; small and large rings. Reactivity of heterocycles: overview; six-membered rings; five-membered rings with one heteroatom; five-membered heterocycles with two or more heteroatoms; small and large rings. Synthesis of heterocycles: overview; monocyclic rings with one heteroatom; monocyclic rings with two or more heteroatoms; bicyclic ring systems without ring junction heteroatoms; tri- and polycyclic ring systems with ring junction heteroatoms. Subject index.

5.161

Sainsbury, Malcolm. *Heterocyclic chemistry.* NY: Wiley-Interscience; [Cambridge, UK]: Royal Society of Chemistry, 2002. 142 p. (Basic concepts in chemistry) 0471281646 pbk QD400 547 2002-725621

Discusses the fundamental chemistry of fully saturated and unsaturated 4-, 5-, and 6-membered heterocycles, but concentrates on those with a single nitrogen, oxygen, or sulfur atom. Treats heterocyclics important in life, medicine, and industry; includes conformation, aromatic stabilization, nomenclature, reaction mechanisms, and synthetic methods.

Chemistry of Natural Products, Including Alkaloids

5.162

The chemistry of natural products, ed. by K. W. Bentley. NY: Interscience. QD415

V. 1: The alkaloids, by K.W. Bentley, 1957; v. 2: Mono and sesquiterpenoids, by P. de Mayo, 1959; v. 3: The higher terpenoids, by P. de Mayo, 1959; v. 4: The natural pigments, by K.W. Bentley, 1960; v. 5: The carbohydrates, by S.F. Dyke, 1960.; v. 6: The chemistry of vitamins, by S.F. Dyke, 1965; v. 7: The alkaloids, Part 2, by K. W. Bentley, 1965.

5.163

Comprehensive natural products chemistry, ed. by Derek Barton, Koji Nakanishi, and Otto Meth-Cohn. Amsterdam; NY: Elsevier, 1999. 9 v. 008042709x (set) QD415 547.7 98-15249

A definitive, scholarly work, the first comprehensive overview of natural products chemistry. Enzymatic reactions emphasizing biosynthetic pathways, mechanisms, and genetics are the main object of this set, rather than traditional synthetic organic chemistry. Genes that encode biosynthesis enzymes are discussed throughout. Some treatment of chemical and spectral determination of structure is included. For the most part, structural formulas are ample, and stereostructure is clearly denoted. V. 9, the collective index, has a formula index gathered from textual material, tables, figures, and schemes and a thoroughly cross-referenced subject index, but no author index. Each volume includes an author index to citations and references, a cross-referenced subject index, a table of abbreviations, and an overview of contents by the volume editor. Each volume repeats introduction, preface, biography of late co-editor Barton, and superb history of natural product chemistry by co-editor Nakanishi. Contents for each volume: v. 1: polyketides and other secondary metabolites, including fatty acids and their derivatives; fatty acids, eicosanoids, macrolides, lignans, flavonoids; v. 2: isoprenoids, including carotenoids and steroids; biosynthesis of mevalonic acid, terpene biosynthesis, protein phrenylation; v. 3: carbohydrates and their derivatives, including tannins, cellulose, and related lignins; glycosidases, glycosyltransferases, deoxysugars, aldolases, celluloses; v. 4: amino acids, peptides, porphyrins, and alkaloids; biosynthesis of alkaloids, biosynthesis of heme, Vitamin B12 biosynthesis, enzymatic synthesis of penicillins; v. 5: enzymes, enzyme mechanisms, proteins, and aspects of NO chemistry; role of hydrogen bonding in enzymatic reactions, keto-enol tautomerism, nucleophilic epoxide openings, ester hydrolysis, phosphatases, reductases, electrophilic alkylations, isomerizations and rearrangements, thymine dimer photochemistry; v. 6: prebiotic chemistry, molecular fossils, nucleosides, and RNA; RNA structure, ribozymes, viroids; v. 7: DNA and aspects of molecular biology; DNA structure, oligonucleotide synthesis, nucleoside analogues, DNA damage, intercalators, DNA-binding peptides; v. 8: miscellaneous natural products including marine natural products, pheromones, plant hormones, and aspects of ecology; plant hormones, plant chemical ecology, pheromones, insect hormones, microbial hormones, marine natural products.

5.164

Dictionary of alkaloids. Editorial board: G.A. Cordell [et al.]; compiled and edited by Ian Southon and John Buckingham. NY: Chapman and Hall, 1989. 2 v., 1,834 p. 0412249103 (set) RS431 615 88-39785

Part of a series that complements *Dictionary of Organic Compounds,* but no overlap in coverage with the main work. Available as part of HEILBRON.

5.165

Dictionary of natural products on CD-ROM. ver. 4:2. London: Chapman and Hall, 1996. 1 CD-ROM. 0412491508

This product, available by subscription, includes more than 80,000 key compounds on the disc. Some 23 fields may be searched, including name, molecular formula, CASRN, type of compound, biological species, and physical properties. Six-months updates. Part of the publisher's CRCnetBase series.

5.166

Hesse, Manfred. *Alkaloids: nature's curse or blessing.* Weinheim; Chichester: Wiley-VCH, 2002. 413 p. 3906390241 QK898.A4 547

Provides an overview of alkaloid structure, properties, and history. Discusses chiroptical properties and synthetic pathways as well as biological and biochemical aspects. Reviews the cultural and historical significance of the most important alkaloid sources.

5.167

Hostettmann, Kurt, A. Marston, and M. Hostettmann. *Preparative chromatography techniques: applications in natural product isolation.* 2nd, completely rev. and enl. ed. Berlin; NY: Springer, 1998. 244 p. 3540624597 QD272.C4 543 98-116711

Provides an up-to-date guide to preparative chromatographic separation of potential and actual bioactive compounds. Emphasizes applications rather than theory. Outlines modern techniques of sample preparation such as supercritical fluid extraction, solvent partition, and solvent phase extraction. Describes preparative planar chromatography, preparative "pressure liquid chromatography," countercurrent chromatography, isolation of macromolecules, and chiral separations.

5.168

Nakanishi, Kōji. *A wandering natural products chemist.* Washington, DC: American Chemical Society, 1991. 230 p. (Profiles, pathways, and dreams) 0841217750; 0841218013 pbk QD22 540 90-45062

Provides an autobiographical look at a prominent organic chemist who has been active in the field for many years.

5.169

The Total synthesis of natural products, ed. by John ApSimon. NY: Wiley, 1973-91. 9 v. (A Wiley-Interscience publication) 0471032514 (v. 1) QD262 547 72-4075

Organic Chemistry

Offers a definitive resource for total synthetic approaches to a variety of natural products. Chapter authors are experts.

Organometallics and Coordination Compounds

5.170

Comprehensive coordination chemistry: the synthesis, reactions, properties and applications of coordination compounds. Ed. in chief, Geoffrey Wilkinson; executive editors Robert D. Gillard and Jon A. McCleverty. Oxford; NY: Pergamon Press, 1987. 7 v., 7,500 p. 0080262325 (set) QD474 541.2 86-12319

V. 1: Theory and background; v. 2: Ligands; v. 3: Main group and early transition metals; v. 4: Middle transition elements; v. 5: Late transition elements; v. 6: Applications; v. 7: Indexes. Coordination chemistry is the chemistry of metal complexes and covers such diverse fields as dyes, color photography, mineral extraction, nuclear fuels, toxicology and medicine. Industries dealing with organic chemicals, pharmaceuticals, petrochemicals, and plastics are all concerned with coordination chemistry.

5.171

Comprehensive coordination chemistry. 2: From biology to nanotechnology; eds. in chief Jon A. McCleverty and Thomas J. Meyer. Oxford: Pergamon, 2003. 10 v., 8,000 p. 0080437486 QD474 541.2242

V. 1: Fundamentals; v. 2: Coordination chemistry of the s, p, and f metals; v. 3: Transition metal groups 3-6; v. 4: Transition metal groups 7 and 8; v. 5: Transition metal groups 9-12; v. 6: From the molecular to the nanoscale: Synthesis and structure; v. 7: From the molecular to the nanoscale: Properties; v. 8: Bio-coordination chemistry; v. 9: Applications of coordination chemistry; v. 10: Cumulative subject index.

5.172

Comprehensive supramolecular chemistry; exec. ed., Jerry L. Atwood ... [et al.]; chairman of the editorial board, Jean-Marie Lehn. NY: Pergamon, 1996. 11 v. 0080406106 (set) QD411 547 96-21927

A valuable and timely account of this newly and rapidly emerging area of chemistry, which is "chemistry beyond the molecule, referring to the organized entities of higher complexity that result from the association of two or more chemical species held together by intermolecular forces." There are 10 topical volumes and a cumulative index: v. 1, Molecular Recognition: Receptors for Cationic Guests; v. 2, Molecular Recognition: Receptors for Molecular Guests; v. 3, Cyclodextrins; v. 4, Supramolecular Reactivity and Transport: Bioorganic Systems; v. 5, Supramolecular Reactivity and Transport: Bioinorganic Systems; v. 6, Solid-State Supramolecular Chemistry: Crystal Engineering; v. 7, Solid-State Supramolecular Chemistry: Two- and Three-Dimensional Inorganic Networks; v. 8, Physical Methods in Supramolecular Chemistry; v. 9, Templating, Self-Assembly and Self-Organization; and v. 10, Supramolecular Technology. Each volume collects authoritative articles by a large variety of contributors.

5.173

Computational organometallic chemistry, ed. by Thomas R. Cundari. NY: Marcel Dekker, 2001. 428 p. 0824704789 QD411 547 2001-28072

Provides a "how-to" approach to fundamentals, methodologies, and dynamics of this area of organometallic chemistry, including classical and molecular mechanics, quantum mechanics, and the most recent hybrid techniques.

5.174

Constable, Edwin C. *Metals and ligand reactivity: an introduction to the organic chemistry of metal complexes.* New rev. and expanded ed. Weinheim; NY: VCH, 1996. 308 p. 3527292780; 3527292772 pbk QD474 541.2242

(1st ed., 1990.) This revision incorporates new material and corrects previous errors. As before, the book describes alternative approaches to ligand reactivity involving coordination complexes rather than organometallic compounds. Explains organic reactions promoted, catalyzed, or initiated by metal ions. References to individual reactions are not given; rather, general review articles and books for further reading are listed. Topics covered include principles of metal-ligand interaction; reactions of coordinated ligands with nucleophiles and electrophiles; oxidation and reduction of coordinated ligands; cyclic and encapsulating ligands, template effects, and supramolecular chemistry. Examples, schemes, figures, and study problems.

5.175

Coordination chemistry of aluminum, ed. by Gregory H. Robinson. NY: VCH, 1993. 234 p. 1560810599; 1560816562 pbk QD474 546 93-12703

Documents many important discoveries of the past decade, and serves as a timely compendium of research on contemporary synthesis and structure in organoaluminum coordination chemistry. Contents includes normal and dative bonding in neutral aluminum compounds; organoaminoalanes: unusual Al-N systems; aqueous coordination chemistry of aluminum; low valent and paramagnetic compounds of aluminum; the chemistry of alkoxides, thiolates, and the heavier Group 6 derivatives of aluminum and gallium; anionic and cationic organoaluminum compounds.

5.176

Davies, Alwyn G. *Organotin chemistry*. Weinheim; NY: VCH, 1997. 327 p. 3527290494 QD412.S7 547 97-194256

Concentrates on organometallic aspects of organotin compounds; includes preparation, structure variety, and chemistry of applications. Accompanying CD-ROM includes more than 2,500 references to the literature, cross-references to relevant sections of the book.

5.177

Dodziuk, Helena. *Introduction to supramolecular chemistry.* Dordrecht; Boston: Kluwer Academic Publishers, 2002. 350 p. 1402002149 QD878 547 2001-50614

A rapidly developing field relating chemistry, biochemistry, physics, and technology that will help in understanding the functioning of living organisms and the origin of life.

5.178

Hegedus, Louis S. *Transition metals in the synthesis of complex organic molecules.* 2nd ed. Sausalito, CA: University Science Books, 1999. 337 p. 1891389041 QD262 547 99-10276

An excellent resource for organometallic chemistry. The first two chapters remain unchanged, and are the strength of the book, introducing electron counting in organometallic compounds and reaction mechanisms of these compounds, respectively. Other chapters treat syntheses of organic compounds through transition metal complex intermediates such as metal hydrides, metal carbon sigma bonds, metal carbonyls, and metal carbene, alkene, alkyne, allyl, and arene compounds. Information is current through 1998. New sections were added to chapter 5, "Bridging Acyl Complexes"; chapter 6, "Metathesis Processes of Electrophilic Carbene Complexes"; and chapter 7, "Metal-Catalyzed Cycloaddition Reactions."

5.179

Houghton, Roy Peter. *Metal complexes in organic chemistry.* Cambridge; NY: Cambridge University Press, 1979. 308 p. (Cambridge texts in chemistry and biochemistry) 0521219922; 0521293316 pbk QD411 547 78-51685

Surveys the principles of metal complexes; organized by reaction types of organic chemistry, including substitutions, additions, etc. Includes reactions in which metal ions are catalysts.

5.180

Coates, Geoffrey Edward ... [et al.]. *Organometallic compounds.* 4th ed. London: Chapman and Hall; NY: J. Wiley: distributed by Halsted Press, 1979- . v. <v. 1, pt. 2> in <1>v. 0412130203 (v. 1, pt. 2) QD411 547 80-458482

V. 1: The main group elements. Primarily a reference work, with a large number of references to the original literature. Forms a good starting point for serious study of organometallic chemistry.

5.181

Handbook of organopalladium chemistry for organic synthesis, ed. by Ei-ichi Negishi; A. de Meijere, assoc. ed. NY: Wiley-Interscience, 2002. 2 v., 3,279 p. 0471315060 QD262 2002-726894

Organized by reaction type. Contents: palladium compounds: stoichiometric preparation, in-situ generation, and physical and chemical properties; palladium-catalyzed reactions involving reductive elimination: cross-coupling; carbopalladation and related reactions; palladium-catalyzed reactions involving nucleophilic attack on ligands; palladium-catalyzed carbonylation and other related reactions involving migratory insertion; palladium-catalyzed hydrogenation and related reduction reactions; palladium-catalyzed oxidation reactions; rearrangement and other reactions catalyzed by palladium; and technological developments in organopalladium chemistry.

ORGANIC ANALYTICAL PROCEDURES

5.182

Blackburn, Stanley. *Amino acids and amines.* Boca Raton, FL: CRC Press, 1983. 2 v. (CRC handbook of chromatography) 0849330645 (v.1); 0849330661 (v. 2) QD79 543 82-9561

Contains sections on detection reagents, principally for paper and thin-layer chromatography; methods of sample preparation; determination of amino acids by liquid and gas chromatography, ion-exchange resins, amino acid enantiomers, and detection of amino acids in automatic analyzers; commercial sources of chromatographic materials; and a list of references.

5.183

Clarke, Hans Thacher. *A handbook of organic analysis: qualitative and quantitative*. 5th ed. Rev. by B. Haynes; with E.C. Brick and G.G. Shone. London: E. Arnold, 1975. 291 p. 0713124601; 0713124903 QD271 547.3 75-330409

Distributed in the U.S. by Crane, Russak & Co., NY. Became a standard work for students, especially in England. Covers much information, especially the extensive tables of properties of compounds and derivatives.

5.184

Crews, Phillip, Jaime Rodríguez, and Marcel Jaspars. *Organic structure analysis*. NY: Oxford University Press, 1998. 552 p. (Topics in organic chemistry) 0195101022 QD272.S6 547 97-31686

Summarizes major techniques in present use: NMR, mass spectroscopy, IR, and optical techniques. Sample spectra are used to illustrate each technique.

5.185

Enthalpies of vaporization of organic compounds: a critical review and data compilation, ed. by Vladimir Majer and Vaclav Svoboda. Oxford: Blackwell Scientific, 1985. 300 p. (IUPAC Chemical Data Series no. 32) 0632015292 QC304 547.1 85-36273

Contents: Introduction; Enthalpy of vaporization—basic relations and major applications; Non-calorimetric methods for determining enthalpies of vaporization and data availability; Calorimetric methods for determining enthalpies of vaporization of pure substances; Critical compilation of enthalpies of vaporization and cohesive energies; Guide to tables.

5.186

Feigl, F., and Vinzenz Anger. *Spot tests in organic analysis*. 7th English ed., completely rev. and enl. Tr. from German by Ralph E. Oesper. NY: Elsevier, 1966. 772 p. QD271 547.34834 65-13235

Companion to the inorganic Feigl volume. Both these were parts of an original, larger work available in German. Small-scale testing procedures are featured. Developments, present state, and prospects of organic spot test analysis; preliminary (exploratory) tests, including those for elements; determination of characteristics for groups; determination of structures and certain types of organic compounds; identification of individual organic compounds; spot tests in the differentiation of isomers and homologous compounds; purity and pharmaceutics; and other topics.

5.187

Feinstein, Karen. *Guide to spectroscopic identification of organic compounds*. Boca Raton, FL: CRC Press, 1995. 124 p. 0849394481 QD272.S6 547.3 94-38923

Guide for interpreting organic spectra, with diagnostic spectroscopic aids, problem-solving algorithms, and other aids. Lists physical properties (boiling and melting point, density, refractive index); supplies CASRN and Beilstein References.

5.188

IUPAC. Commission on Electrochemical Data. *Dissociation constants of organic acids in aqueous solution*, by G. Kortüm, Gustav, W. Vogel, and K. Andrussow. London: Butterworth, 1961. 1 v., various pagings. (Reprinted from *Pure and Applied Chemistry* v. 1, no. 2-3.)

5.189

IUPAC. Commission on Equilibrium Data. *Ionisation constants of organic acids in aqueous solution*, ed. by E.P. Serjeant and Boyd Dempsey. Oxford; NY: Pergamon Press, 1979. 989 p. (IUPAC Chemical Data Series no. 23). 0080223397 QD561 547 78-40988

Supplement to IUPAC Commission ... *Dissociation Constants of Organic Acids in Aqueous Solution*, 1961, by Kortüm et al. Arranged numerically in ascending order of number of carbons, by molecular formula.

5.190

Pedley, J.B., R.D. Naylor, and S.P. Kirby. *Thermochemical data of organic compounds*. 2nd ed. London; NY: Chapman and Hall, 1986. 792 p. 0412271001 QD511.8 547.1 85-13206

Revised edition of Sussex-N.P.L. *Computer Analysed Thermochemical Data*, 1977. Used to predict which chemical processes are thermodynamically feasible; 3,000 compounds listed. Excludes radicals, ions, organometallics, and multicomponent systems. Name and CASRN indexes.

5.191

Perrin, D.D. *Dissociation constants of organic bases in aqueous solution.* London: Butterworths, 1965. 473 p. QD273 547.134 65-2983 (Published as a supplement to *Pure and Applied Chemistry.*)

At head of title: International Union of Pure and Applied Chemistry, Analytical Chemistry Division, Commission on Electroanalytical Chemistry. Contains data on pK and ionization constants.

5.192

The systematic identification of organic compounds, by R.L. Shriner ... [et al.]. 7th ed. NY: J. Wiley, 1998. 669 p. 0471597481 QD261 547 97-5545

(5th ed., 1964, had as authors Shriner, R.C. Fuson, and D.Y. Curtin; 6th ed., 1980) Discusses identification of unknowns, preliminary examination, determination of physical properties, qualitative analysis for elements, solubility classes, application of classification tests, spectroscopic determination of functional groups, preparation of derivatives, tables of derivatives, separation of mixtures, and discussion of solution of structural problems.

5.193

Timmermans, Jean. *Physico-chemical constants of pure organic compounds.* NY: Elsevier, 1950-65. 693 p. (2 v.) QD291 547 50-9668

Bibliography, p. 631-660. At head of title: International Union of Pure and Applied Chemistry. Lists boiling point, melting point, density, viscosity, surface tension, refractive index, critical constants, specific heat, saturated vapor pressure, dielectric constant, specific rotatory power, and magnetic susceptibility. Arranged by class of compound; formula and subject indexes.

LABORATORY PROCEDURE MANUALS

5.194

Armarego, W.L.F., and D.D. Perrin. *Purification of laboratory chemicals.* 4th ed. Oxford; Boston:

Butterworth Heinemann, 1996. 529 p. 0750628391 TP156 542 97-109714

An indispensable reference source that provides concise information about purification methods for chemicals, mostly organic, used in laboratories. Approximately 4,000 organic substances, as well as classes of organic compounds, are included. Arrangement is alphabetic by specific substance and class of compound. Common names are used, with liberal cross-references. Information provided includes physical properties, such as melting and boiling points; molecular weight; density (occasionally); refractive indexes; and purification methods (ranging from brief statements about crystallization to extensive experimental details). Occasionally a journal reference is listed with additional experimental information. (Information from 3rd ed.) (Perrin's name appears first on earlier editions.)

5.195

Fieser, Louis F., and Kenneth L. Williamson. *Organic experiments.* 8th ed. Boston: Houghton Mifflin Co., 1998. 644 p. 0395865190 QD261 547 97-72467

Includes apparatus, operations, weights and measures, distillation, fractional distillation, melting point determination, crystallization, extraction, steam distillation, list of various preparations.

5.196

Landgrebe, John A. *Theory and practice in the organic laboratory; with microscale and standard scale experiments.* 4th ed. Pacific Grove, CA: Brooks/Cole Pub., 1993. 586 p. 053416854x QD261 547 92-32949

Includes 52 microscale and standard scale procedures and experiments. Covers safety and hazardous waste disposal; lab techniques for handling, synthesis, separation, and purification of organic compounds; includes spectroscopic methods for identification.

5.197

Menger, Fredric M., and Leon Mandell. *Electronic interpretation of organic chemistry: a problems-oriented text.* NY: Plenum Press, 1980. 216 p. 030640379x; 0306403919 pbk QD502 547 79-21718

Stresses the "feel" for the chemistry rather than memorization. Offers a stepwise explanation of underlying principles of reactions or mechanisms to explain observed results. Chapters treat basic principles (bonding, acid-base theory, etc.), solved problems, problems involving reactive intermediates (carbocations, carbenes, radicals, etc.), and molecular orbital theory.

5.198

Zubrick, James W. *The organic chem lab survival manual: a student's guide to techniques.* 4th ed. NY: Wiley, 1997. 382 p. 0471129488 pbk QD261 547 96-35813

Describes instruments and techniques used in an organic laboratory. Contains directions for finding organic chemical laboratory information on the Internet.

Problem Manuals

5.199

Ghiron, Chiara, and Russell J. Thomas. *Exercises in synthetic organic chemistry.* Oxford; NY: Oxford University Press, 1997. 123 p. 0198559445; 0198559437 pbk QD262 547 96-47593

Each of the 82 synthetic exercises in this book offers the total synthesis of a natural product or related system, with key reagents or intermediates missing; readers are to supply the missing items, and meet with others to discuss the solutions.

5.200

Meislich, Herbert, Howard Nechamkin, and Jacob Sharefkin. *Schaum's outline of theory and problems of organic chemistry.* 3rd ed. NY: McGraw-Hill, 1999. (Schaum's outline series) 007134165x QD257 547 99-28581

Treats structure and properties of organic compounds, bonding and molecular structure, chemical reactivity and organic reactions, alkanes, stereochemistry, alkenes, alkyl halides, alkynes and dienes, cyclic hydrocarbons, benzene and polynuclear aromatic compounds, aromatic substitution, arenes, spectroscopy and structure, alcohols and thiols, ethers, epoxides, glycols, thioethers, carbonyl compounds (aldehydes and ketones), carboxylic acids and derivatives, carbanion-enolates and enols, amines, phenolic compounds, and aromatic heterocyclic compounds.

5.201

Patrick, Graham L. *Instant notes organic chemistry.* Oxford: BIOS Scientific Publishers, 2000. 315 p. (The Instant notes series) 1859961584 547

Provides concise and comprehensive discussion of organic chemistry for undergraduates. Treats structure and bonding; alkanes and cycloalkanes; functional groups; stereochemistry; alkenes and alkynes; aromatic chemistry; aldehydes and ketones; carboxylic acids and carboxylic acid derivatives; alkyl halides; alcohols, phenols, and thiols; ethers, epoxides, and thioethers; and amines and nitriles.

Chapter 6

Inorganic Chemistry

In this part of our journey through the chemical literature, the smaller area of inorganic chemistry will be surveyed. Because there are limited means of combing atoms that exclude organic carbon, the result is a more "tidy" (my words as a high school chemistry student) area of chemistry (one which I vowed to study in preference to the "untidy" area of organic chemistry. I did not take my own advice.). As in earlier parts of this lengthy examination of the chemical literature, works that were discussed or mentioned in previous chapters will not be included. An effort has been made to select materials for this section that will be useful to academic and student chemists; there are other works that would certainly be suitable for inclusion, but may not be the best place to go for information on account of their advanced treatment of the subject. Because chemistry is the largest body of knowledge in the world, there are a multitude of places in which to look for answers to chemical questions.

GENERAL SURVEYS

6.001

Mackay, K.M., R.A. Mackay, and W. Henderson. *Introduction to modern inorganic chemistry*, 5th ed. London: Blackie, 1989; reprinted in 1998 by Stanley Thornes (Publishers) Ltd. 468 p. 0748739831 QD151.2 546 88-55765 (4th ed.)

Provides the fundamental chemistry needed to develop a thorough understanding of inorganic chemistry. Offers a broad overview; addresses biological, medicinal, and environmental aspects; the periodic table is used as the basis of a systematic survey of the elements.

6.002

Mark, James E., Harry R. Allcock, and Robert West. *Inorganic polymers.* Englewood Cliffs, NJ: Prentice-Hall, 1992. 272 p. (Prentice Hall advanced reference series; Physical and life sciences) 0134658817 QD196 546 91-9936

The book contains three main chapters on polyphosphazenes, polysiloxanes, and polysilanes with a brief discussion of basic chemistry and structure-property relationships for the three polymer types. Some miscellaneous inorganic polymers, covering germanium, sulfur and selenium, and boron and aluminum, are briefly described in a terminal chapter.

6.003

Mingos, D.M.P. *Essential trends in inorganic chemistry.* Oxford; NY: Oxford University Press, 1998. 392 p. 0198501099; 0198501080 pbk QD467 546 97-35359

Shows exactly how a chemist uses the periodic table to organize and process information.

6.004

Moeller, Therald. *Inorganic chemistry, a modern introduction.* NY: Wiley, 1982. Reprinted by R.E. Krieger in 1990. 846 p. 0471612308 QD151.2.M63 546 81-16455

Treats atomic structure and related properties; periodic table and periodic relationships; chemical bond; molecular structure and stereochemistry; coordination chemistry; inorganic chemical reactions.

6.005

Sidgwick, Nevil Vincent. *Chemical elements and their compounds.* Oxford: Clarendon Press, 1950. 2 v., 1,703 p. QD466 546 50-4087

This is an old reference, but it contains good material and is very useful.

NOMENCLATURE

6.006

Block, B. Peter, Warren H. Powell, and W. Conard Fernelius. *Inorganic chemical nomenclature: principles and practices.* Washington, DC: American Chemical Society, 1990. 210 p. (ACS professional reference book)

0841216975; 0841216983 pbk QD149 546 90-760

Covers in 16 chapters: language and nomenclature; fundamental principles; homoatomic species; heteroatomic species: general principles; additive nomenclature; polynuclear coordination entities; polymeric inorganic species; acids, bases, and their derivatives; substitutive nomenclature for covalent inorganic compounds; chains and rings; boron compounds; organometallic compounds; addition compounds; nonstoichiometric species; isotopically modified species; stereochemical relationships. Appendix. Bibliographies.

6.007

Commission on the Nomenclature of Inorganic Chemistry. *Nomenclature of inorganic chemistry: recommendations, 1990*. Ed. by G.J. Leigh. Oxford; Boston: Blackwell Scientific Publ., 1990. 320 p. 0632024941 (pt. 1) QD149 546 96-163414

Issued by the Commission, International Union of Pure and Applied Chemistry (IUPAC). Contents: General aims, functions, and methods of chemical nomenclature; Grammar; Elements, atoms, and groups of atoms; Formulae; Names based on stoichiometry; Solids; Neutral molecular compounds; Names for ions, substituent groups and radicals, and salts; Oxoacids and derived anions; Coordination compounds; Boron hydrides and related compounds; Tables; Appendix; Index.

6.008

International Union of Pure and Applied Chemistry. Commission on the Nomenclature of Inorganic Chemistry. *Nomenclature of inorganic chemistry. II, Recommendations 2000*, issued by the Commission ... and ed. by J.A. McCleverty and N.G. Connelly. [3rd ed.] Cambridge, UK: Royal Society of Chemistry, 2001. 130 p. 0854044876 QD149 546.014

This volume is Part 2 of the rev. ed. of *Nomenclature of Inorganic Chemistry: Definitive Rules 1970*, 2nd ed., 1971. The first volume (Part 1) appeared in 1990. Discusses polyanions; isotopically modified inorganic compounds; metal complexes of tetrapyrroles; hydrides of nitrogen and derived cations, anions and ligands; inorganic chain and ring compounds; graphite intercalation compounds; regular single-strand and quasi single-strand inorganic and coordination polymers.

ENCYCLOPEDIA

6.009

Encyclopedia of inorganic chemistry; ed. in chief, R. Bruce King. Chichester; NY: Wiley, 1994. 8 v., 4,819 p. 0471936200 (set) QD148 546 94-19739

Features reviews of the chemistry of all the elements. The entries are listed alphabetically and examine the six major areas of inorganic chemistry: Physical and theoretical methods; Organometallic chemistry; Main group elements; Solid state chemistry; Transition metals; Bioinorganic chemistry. The encyclopedia contains a contents listing for each of 260 main articles; 850 definition entries on important concepts in inorganic chemistry; an introduction to each article, providing an entry to the topic; extensive illustrations; including line drawings, photographs, structural formulas, and equations and schemes; and comprehensive cross-referencing with lists of related articles for further reading and extensive reference lists.

HANDBOOKS AND TABLES OF DATA

6.010

Dictionary of inorganic compounds. Exec. ed., J.E. Macintyre. London; NY: Chapman and Hall, 1992. 5 v., 5,400 p. 0412301202 QD148 546 92-250067 Updated by annual supplements.

V. 1: Ac-C10; v.2: C11-C45; v.3: C46-Zr; v. 4: Index of commonly occurring structural types, name index, CAS Registry Number index; v. 5: Element index. This reference was compiled under the guidance of an international panel of experts and contains physical, structural, and bibliographic information on approximately 42,000 compounds. Each entry contains bibliographic references that are extensively labeled to allow the user rapid access to the primary literature, and the entries are organized alphabetically by formula according to the Hill convention.

6.011

Gmelins Handbuch der anorganischen Chemie. 8. Aufl. Hrsg. von der Deutschen Chemischen Gesellschaft, bearb. von R. J. Meyer, unter beratender Mitwirkung von Franz Peters. Berlin: Springer Verlag, 1924-<1990 >. 590 v. Volumes since 1980 in English. QD151 546 25-1383

Assembles and systematically classifies research on the inorganic chemistry of the elements. The Gmelin system uses a System number for each element; within each subject classification, the arrangement is in terms of inorganic and physical chemistry, with further subdivisions into analytical chemistry; atomic physics; ore dressing; chemical technology; iron and steel; electrochemistry; geochemistry; history of chemistry; colloid chemistry; coordination chemistry; corrosion and passivity; crystallography; geology; metallography; metallurgy; mineralogy; nonferrous metals; physical properties of elements, compounds, and alloys; toxicity and hazards; and economic and statistical data.

6.012
Handbook of inorganic compounds, ed. by Dale L. Perry and Sidney L. Phillips. Boca Raton, FL: CRC Press, 1995. 1 v., various pagings 0849386713 QD155.5 546 95-15997

Contains physical and crystallographic data, nomenclature, and references for about 4,000 inorganic materials, selected on the basis of frequency of use in laboratories, mention in recent publications, and other recommendations. Includes names, synonyms, molecular formulas, CASRNs, and physical property data.

6.013
Martell, Arthur E., and Robert M. Smith. *Critical stability constants*. NY: Plenum Press; 1974-89. 6 v. 0306352117 (v. 1); 0306352125 QD503 541 74-10610

Contains critically evaluated metal complex equilibrium constants and corresponding enthalpy and entropy values. Data are arranged by ligand: v. 1: Amino acids; v. 2: Amines; v. 3: Other organic ligands; v. 4: Inorganic complexes; v. 5: First supplement; v. 6: Second supplement.

6.014
The NBS tables of chemical thermodynamic properties: selected values for inorganic and C1 and C2 organic substances in SI units, by Donald D. Wagman [et al.]. NY; Washington, DC: American Chemical Society and the American Institute of Physics for the National Bureau of Standards, 1982. 392 p. (JPCRD-NBS, 11, Suppl. 2; Revision of NBS Technical Note 270: Selected values of chemical thermodynamic properties by D.D. Wagman) 0883184176 QD511.8 83-70427

Recommended values for these compounds, including enthalpy of formation, entropy, heat capacity, and others, in SI units.

6.015
Patnaik, Pradyot. *Handbook of inorganic chemicals*. NY: McGraw-Hill, 2003. 1,086 p. (McGraw-Hill handbooks) 0070494398 QD155.5 546 2002-29526

Lists synonyms, molecular weight, formulas and structure, CASRN, occurrence, uses and applications, physical and chemical properties, preparation, reactions, and health concerns.

6.016
Perrin, Douglas Dalzell, compiler. *Ionisation constants of inorganic acids and bases in aqueous solution*. 2nd ed. Oxford; NY: Pergamon Press, 1982. 180 p. (IUPAC chemical data series, no. 29) 0080292143 QD561 541.3 82-16524

Bibliography, p. 139-180. (Revised edition of *Dissociation Constants of Inorganic Acids and Bases in Aqueous Solution*, 1969.) At head of title: International Union of Pure and Applied Chemistry, Analytical Chemistry Division, Commission on Equilibrium Data.

6.017
Thermochemical properties of inorganic substances, ed. by O. Knacke, O. Kubaschewski, and K. Hesselmann. 2nd ed. Berlin; Heidelberg; NY: Springer-Verlag, 1991. 2 v., 129, 2,412 p. 3540540148; 0387540148; 3514003637 QD511.8 541.3 91-25087

Includes bibliographical references (v. 1, p. [113]-129) and index. Contains tables with comprehensive data ranging from entropy, enthalpy, vapor pressure, chemical equilibria, to thermal decomposition, stability and metastability, and thermochemical energy balance.

ADVANCES AND REVIEWS

6.018
Advances in inorganic chemistry. Orlando, FL: Academic Press, 1987- . Irregular. ISSN 0898-8838 QD151 546 88-649836

Presents timely and informative summaries of current progress in numerous areas within inorganic chemistry, ranging from bioinorganic to solid state (v. 52 published in 2001).

6.019

Advances in inorganic chemistry and radio-chemistry, ed. by H.J. Eméleus and A.G. Sharpe, 1959-1986. NY: Academic Press, v. 1-30, 1959-86. ISSN 0065-2792 QD151 546.082 59-7692

Contains about six review articles of about fifty pages each.

6.020

Progress in inorganic chemistry, ed. by F. Albert Cotton. NY: Wiley-Interscience. v. 1- , 1959- . Annual. ISSN 0079-6379 QD151 546.082 59-13035

Each volume contains a cumulative index to preceding volumes. Each volume contains about six review articles of about fifty pages each.

COMPILATIONS AND TREATISES

6.021

Comprehensive inorganic chemistry. Editorial board: J. C. Bailar, Jr.; H.J. Eméleus, Sir Ronald Nyholm [and] A.F. Trotman-Dickenson (exec. ed.) [Oxford]: Pergamon, [1973]; distributed by Compendium Publishers. 5 v. 008017275x (Compendium) QD151.2 546 77-189736

V. 1, H, Noble gases, Group IA, Group IIA, Group IIIB, C, and Si. V. 2: Ge, Sn, Pb, Group VB, Group VIB, Group VIIB; v. 3: Group IB, Group IIB, Group IIIA, Group IVA, Group VA, Group VIA, Group VIIA, Group VIII; v. 4: Lanthanides, transition metal compounds; v. 5: Actinides; Master index.

6.022

Cotton, F. Albert. *Advanced inorganic chemistry*, 6th ed. NY: Wiley, 1999. 1,355 p. 0471199575 QD151.2 546 98-8020

This work describes, by family of inorganic elements, physical methods, properties, bonding. The new edition covers metal cluster compounds, stereochemistry, molecular symmetry, cage and coordination polyhedral structures; stereochemical non-rigidity and fluxionality; metal-metal bonding and metal cluster compounds; bioinorganic chemistry of iron, cobalt and some other metals; a more detailed treatment of complexes with CO, CNR_2, N_2, O_2, NO, complexes; and homogeneous catalysis by low-valent metal complexes. (Revision of 5th ed., 1988, by Cotton and Geoffrey W. Wilkinson.)

6.023

Cox, P.A. *The elements on earth: inorganic chemistry in the environment*. Oxford; NY: Oxford University Press, 1995. 287 p. 0198562411; 0198559038 pbk QD31.2 546 94-42633

A broad overview of geochemistry: structure of the Earth, its chemical composition, the processes that control elemental distribution, and human impacts on the environment. Numerous figures show elemental abundances in the crust, whole Earth, seawater, and solar system; specific topics include elemental abundances, properties, isotopes, oxidation state, coordination number, aqueous solubility, biological activity, geochemical cycling, usage, extraction, and reserves as well as atmospheric and environmental chemistry. The second part discusses occurrence and chemical behavior of each named element, including important elements such as iron, oxygen, carbon, and hydrogen.

6.024

Cox, P.A. *Inorganic chemistry*. Oxford: BIOS; NY: Springer, 2000. (The instant notes series) 0387916040 pbk 1859961630 QD153.5 546 00-39829

Provides concise and comprehensive treatment to core information in inorganic chemistry. Treats atomic structure, chemical trends, structure and bonding in molecules and solids, solution chemistry, nonmetal chemistry, non-transition metal chemistry, transition metal chemistry, lanthanides and actinides, and environmental, biological, and industrial matters.

6.025

Elements of the p block, ed. by Charlie Harding, Rob Janes, and David Johnson. Cambridge, UK: Royal Society of Chemistry, 2002. 305 p. 2 CD-ROMs (The molecular world) 0854046909 QD466 546

Treats the chemistry of the p-block elements and hydrogen; introduces chemical bonding, oxidation numbers, bond strengths, dipole moments, and intermolecular forces; mentions the chemistry of carbon nanotubes.

6.026

Greenwood, Norman Neill, and Alan Earnshaw. *Chemistry of the elements*. 2nd ed. NY: Pergamon, 1997. 1,340 p. 0750633654 pbk QD466 546 97-36336

(Information taken from 1st ed.) More than 3,000 illustrations with 1,500 literature references. Covers inorganic chemistry of the elements, analytical, theoretical, industrial organometallic, bioinorganic, and other areas of chemistry that apply. The chemistry of the elements is discussed within the context of an underlying theoretical framework, but at times the chemical facts are emphasized. It is written for graduate and undergraduate students.

6.027

Henderson, William. *Main group chemistry.* NY: Wiley-Interscience; [Cambridge, England]: Royal Society of Chemistry, 2002. 196 p. (Basic concepts in chemistry) 0471224782 pbk QD151.3 546 2002-6102

Discusses the chemistry of elements of the s- and p-block, together with a short chapter on zinc, cadmium, and mercury.

6.028

Heslop, R.B., and K. Jones. *Inorganic chemistry: a guide to advanced study.* Amsterdam; NY: Elsevier Scientific Pub. Co., 1976. 830 pp. 0444414266 (American Elsevier) QD151.2 546 76-364266

This text is the successor to R.B. Heslop and P.L. Robinson, 1967. Surveys elements by periodic table group arrangement; physical methods of structure determination, bonding, and molecular and orbital symmetry, exclusive use of SI units with the appropriate alterations in physical formulae when the international system makes them necessary.

6.029

Jones, Chris J. *d- and f-block chemistry.* NY: J. Wiley, 2002. 175 p. (Basic concepts in chemistry) 0471224766 QD172.T6 2002-9098

Introduces the principles underlying the chemistry of the d- and f-block metals. Describes origins, uses, and importance; structure, bonding, chemical thermodynamics, and spectroscopy.

6.030

Kettle, S.F.A. *Physical inorganic chemistry: a coordination chemistry approach.* Oxford; NY: Oxford University Press, 1998. 490 p. 0198504055; 0198504047 pbk QD475 541.2 98-205523

Originally published by Spektrum in 1996. User-friendly introduction to important topics and techniques in coordination and physical inorganic chemistry. Provides an account of traditional topics and introduces significant new research. Theoretical, spectroscopic, and physicochemical techniques are included.

6.031

Mellor, Joseph William. *A comprehensive treatise on inorganic and theoretical chemistry.* London, NY: Longmans, 1922-37. 16 v., suppl. QD31 540.2 22-7753

Provides various basic chemical techniques (reactions, syntheses, analytical techniques) via established experimental methods. Proceeds through the Periodic Table with most common elements (H, O) first. V. 16 contains general indexes.

6.032

Put, Paul J. van der. *The inorganic chemistry of materials: how to make things out of elements.* NY: Plenum Press, 1998. 391 p. 0306457318 QD151.2 546 98-33993

Collects information important for materials synthesis but not found in standard inorganic books, in a guide to the why, how, and what-for of materials chemistry. Chapters include bonding theory, inorganic molecules, solid state chemistry, surface chemistry, morphology changes, synthesis, and materials design. Physical chemistry information is in the final chapter. Each chapter includes references from books, monographs, review articles, and some recent primary research articles.

6.033

Remy, Heinrich. *Treatise on inorganic chemistry;* translated by J.S. Anderson; ed. by J. Kleinberg. Amsterdam; NY: Elsevier Pub. Co., 1956. 2 v., 1,714 p. QD151 546 57-8603

V. 1: Introduction and main groups of the periodic table. V. 2: Subgroups of the periodic table and general topics. Comprehensive and systematic discussion of chemistry and physics of inorganic compounds arranged according to the periodic table. Gives the history, occurrences, preparation, properties, and uses of the compounds discussed.

6.034

Sneed, Mayce Cannon, ed. *Comprehensive inorganic chemistry,* ed. by M. Cannon Sneed, J. Lewis Maynard, and Robert C. Brasted. NY: Van Nostrand, 1953. V. 1 and 5 reprinted by R.E. Krieger Pub. Co., 1972. 11 v. QD151 546 53-8775

V. 1: Principles of atomic and molecular structure, by W.N. Lipscomb; Theoretical and applied nuclear chemistry; The actinide series. V. 2: Copper, silver, and gold. V. 3: The halogens. V. 4: Zinc, cadmium, and mercury; Scandium, yttrium, and the lanthanide series. V. 5: Nitrogen, phosphorus, arsenic, antimony, and bismuth; Nonaqueous chemistry. V. 6: The alkali metals; Hydrogen and its isotopes. V. 7: The elements and compounds of Group IVA. V. 8: Sulfur, selenium, tellurium, polonium, and oxygen. Provides the history, occurrence, preparation, physical and chemical properties, structure, uses, and the physiological action of compounds listed.

6.035

Swaddle, Thomas Wilson. *Inorganic chemistry: an industrial and environmental perspective*. San Diego: Academic Press, 1997. 482 p. 0126785503 QD151.5 546 97-1552

A unique array of topics not available in any other single volume. Interspersed throughout are discussions of the theory and practical applications of inorganic chemistry, covering an exceptional range of industrial and environmental topics. Discusses thermodynamics, kinetics, and the solid state, and describes the chemistry of inorganic polymers, semiconductors, metallurgy, catalysis, ion-exchange materials, glasses, atmospheric chemistry and pollution control, agriculture, cement, water conditioning, fuel cells, corrosion of metals, and the solid state (gels, electronic materials, superconductive materials, and fullerenes).

STRUCTURE AND PROPERTIES

6.036

Barrett, Jack. *Structure and bonding*. NY: Wiley-Interscience; Cambridge, UK: Royal Society of Chemistry, 2002. 181 p. (Basic concepts in chemistry) 0471224790 QD461 541.2 2002-284913

Explains factors governing covalent bond formation, lengths and strengths of bonds, and molecular shapes. Discusses periodic table principles, periodicity of various atomic properties, and those properties relevant to chemical bonding.

6.037

Bartecki, Adam, and John Burgess. *The colour of*

metal compounds. Amsterdam: Gordon & Breach, 2000. 232 p. 9056992503 QD113 546

One of the fundamental descriptions of any chemical substance is its color. This book discusses the origin and analysis of color in metal compounds, including chapters on analyzing color, the origin of color in metal complexes (even lanthanide complexes), the variations in color as ligands or solvent systems change, and the colors of solids. An expanded English version of a book originally published in Polish in 1993. Updates to the English version include a chapter on using color in teaching inorganic chemistry.

6.038

Comba, Peter, and Trevor W. Hambley. *Molecular modeling of inorganic compounds*. NY: VCH, 1995. 197 p. 3527290761 QD480 541.2 95-24487

This extremely useful book will determine if molecular modeling using the molecular mechanics (MM) approach can solve chemical problems of interest. After a basic outline of MM, a section discusses the mathematical functions that model through-bond and through-space forces and empirical data that describe the "force field" determining the favored structure of the molecule. Most of the book comprehensively reviews systems for which MM has generated useful information about inorganic systems. Structural predictions, metal ion selectivity, spectroscopy, electron transfer, and other topics are discussed primarily for coordination compounds but also for organometallic and bio-inorganic systems. Closing chapters are an instruction manual on how to design a study of an experimental system. Appendix lists commercially available programs with references.

6.039

Ebsworth, E.A.V., David W.H. Rankin, and Stephen Cradock. *Structural methods in inorganic chemistry*, 2nd ed. Boca Raton, FL: CRC Press, 1991. 510 p. 063202965x; 0632029633 pbk QD95 541.2 91-12998

Originally published by Oxford: Blackwell, 1991. Topics addressed include the timescales of physical methods, the relative advantages and disadvantages of those methods, nuclear magnetic resonance spectroscopy, and rotational and vibrational spectroscopy. The book utilizes research examples to illustrate the use of the techniques in real research publications.

6.040

Figgis, Brian N., and Michael A. Hitchman. *Ligand*

field theory and its applications. NY: Wiley-VCH, 2000. 354 p. (Special topics in inorganic chemistry) 0471317764 QD475 541.2 99-28986

The previous version, Figgis's *Introduction to Ligand Fields*, was a classic. The updated book contains a wealth of new material; most chapters have been completely rewritten, and many new topics have been added. Chapters discuss the concept and scope of ligand field theory as well as some ideas of symmetry and physical properties affected by ligand field theory; quantitative aspects of crystal field theory; angular overlap model of the various geometric structures of metal complexes and calculations based on the ligand field split of the d-orbitals; effect of different ligand fields on the energy levels of the metal ions; thermodynamic effects caused by ligand fields; electronic spectra, magnetic properties, electron paramagnetic resonance spectra, and actinide complexes.

6.041

Hyde, Bruce G., and Sten Andersson. *Inorganic crystal structures.* NY: Wiley, 1989. 430 p. (A Wiley-Interscience publication) 0471628972 QD921 548 88-5492

Bibliographies and indexes. Offers a systematic description of nonmolecular structures, including metals, metal alloys, intermetallics, borides, carbides, minerals, and other inorganics. Demonstrates that crystal structures are based on a few simple structures, showing the number of entities to be remembered are not great.

6.042

Inorganic electronic structure and spectroscopy, ed. by Edward I. Solomon and A.B.P. Lever. NY: Wiley, 1999. 2 v. (A Wiley-Interscience publication) 0471326836 (set); 0471154067 (v. 1); 0471326828 (v. 2) QD95 543 98-38180

Offers excellent coverage of methods and applications of spectroscopy to inorganic chemistry. Most references are current through 1998 with a few in 1999. V. 1 treats ligand field theory as well as ground state, excited state, and high-energy spectroscopic methods, including electron paramagnetic resonance, Mossbauer, luminescence, laser, polarized absorption, infrared and Raman, photoelectron, and X-ray. Calculations include ab initio and density functional methods. V. 2 includes applications, such as bioinorganic chemistry, electron transfer, mixed valence compounds, electrochemistry, and photochemistry; and case studies, including inorganic compounds with metal-metal bonds, metal carbonyls, metallocenes, metal nitrosyls, heme sites, spin crossover compounds, and magnetic materials.

6.043

Jean, Yves and François Volatron. *An introduction to molecular orbitals*; transl. and ed. by Jeremy Burdett. NY: Oxford University Press, 1993. 337 p. 0195069188 QD461 514.2 92-45676

Presents a simplified treatment of molecular orbital theory; includes an introduction to atomic and electronic structure, develops molecular orbital and electronic structure, introduces geometry and reactivity of molecules, and treats molecular orbital, structure, and reactivity problems. Suitable for advanced inorganic studies.

6.044

Kosuge, Kōji. *Chemistry of non-stoichiometric compounds.* Oxford; NY: Oxford University Press, 1994. 262 p. (Oxford science publications) 0198555555 QD921 548 93-21864

Describes these compounds, based on statistical thermodynamics and structural inorganic chemistry. Continued progress in materials science and ceramics requires clear understanding of the chemical and physical properties of inorganic compounds. Provides excellent theoretical and experimental bases for understanding inorganic properties, and gives examples of the practical use of selected inorganics.

6.045

Sanderson, Robert Thomas. *Chemical periodicity.* NY: Reinhold Publ. Corp., 1960. 330 p. (Reinhold physical and inorganic chemistry textbook series) QD467 541.901 60-11081

Presents an unusual treatment of reactivity, structure, and shapes of compounds in relations to the place of the elements on the periodic table.

6.046

Zelewsky, Alexander von. *Stereochemistry of coordination compounds.* Chichester, England; NY: Wiley, 1996. 354 p. (Inorganic chemistry) 0471950572; 047195599x pbk QD474 541.2 95-19754

Considers the stereochemistry of coordination compounds from a topographic or qualitative perspective, and treats the principles of stereochemistry carefully and masterfully. Quality illustrations in good number; these are available on the Internet. Because major areas of coordination stereochemistry are treated as well as applications to polynuclear species, chemical reactions, and catalysis, an advanced knowledge of chemistry is assumed of the reader. Extensive chapter references.

SYNTHETIC METHODS

6.047

Inorganic syntheses. 1st ed., Harold Simmons Booth. NY: McGraw-Hill Book Co., 1939- . Irregular. ISSN 0073-8077 QD156 541 39-23015

V. 1 (1939)-30 (1995), 1 v.(?) V. 4 appeared in 1953; v. 10 in 1967; a volume is issued about every two years at present. This work includes checked and tested methods of preparing important inorganic compounds. All methods are checked before publication. Chapters are arranged according to the periodic chart (v. 32 published in 1998).

6.048

Walton, Harold Frederic. *Inorganic preparations: a laboratory manual*. NY: Prentice-Hall, 1948. 183 p. (Prentice-Hall chemistry series) QD155 546.072 49-119

Old but still quite useful. Treats equipment, crystallization, filtration, washing, drying, ionic and covalent compounds, coordination compounds and complex ions, oxidation and reduction, electrolytic preparations. Offers a list of 52 typical preparations.

REACTIONS

6.049

Basolo, Fred, and Ralph G. Pearson. *Mechanisms of inorganic reactions: a study of metal complexes in solution*. 2nd ed. NY: Wiley, 1967. 701 p. QD171 546.3 66-28755

Contents: Introduction; The theory of the coordinate bond; Substitution reactions of octahedral complexes; Stereochemical changes in octahedral complexes; Substitution reactions of square-planar complexes; Oxidation-reduction reactions; Reactions of transition metal organometallics; Metal ion catalysis; Photochemistry.

6.050

Encyclopedia of chemical reactions, compiled and ed. by Carl Alfred Jacobson. NY: Reinhold Pub. Corp., 1946-1959. 8 v. QD155 546.1 46-822

V. 5 compiled and ed. by C.A. Jacobson with the assistance of Clifford A. Hampel and Elbert C. Weaver; v. 6 compiled by C.A. Jacobson, ed. by Clifford A. Hampel. Addenda in v. 8. This work is alphabetic by elemental symbol, with journal references for each reaction. Tries to list all known inorganic reactions. Certain features of an index are combined with the informational content of an abstract, and references to the journal literature are given for each reaction.

6.051

Gould, Edwin S. *Inorganic reactions and structure*. rev. ed. NY: Holt, Rinehart and Winston, 1962. 513 p. QD31 546 62-9519

This text is devoted to the study of reactions as related to the structure of compounds.

6.052

Inorganic reactions and methods, ed. by J.J. Zuckerman; subject index ed. by A.P. Hagen. Deerfield Beach, FL: VCH Publishers, 1986-c1999. 18 v. 0895732505 (set) QD501 541.3 85-15627

V. 1-2: The formation of bonds to hydrogen (pt. 1-2); v.3-4: The formation of bonds to halogens (pt. 1-2); v.5: Formation of bonds to Group VIB (O, S, Se, Te, Po) (pt. 1); v. 6: Formation of bonds to O, S, Se, Te, Po (pt. 2); v. 7-8: The formation of bonds to C, Si, Ge, Sn, Pb (Pt.1); v. 10-11; v. 12A-12B: The formation of bonds to elements of Group IVB. (C, Si, Ge, Sn, Pb)(pt.4); v. 13; v. 14: Formation of bonds to transition and inner-transition metals; v.15: Electron-transfer and electrochemical reactions; photochemical and other energized reactions; v.16: Reactions catalyzed by inorganic compounds. V. 17: Oligomerization and polymerization formation of intercalation compounds. V. 18: Formation of ceramics. Two index volumes (1999). The reference work compiles the entire synthetic literature currently available. Editorial advisors include the three Nobel laureates, E. O. Fischer, H. Taupe, and Sir G. Wilkinson. The work includes both the classical chemistry of the elements and the frontiers of today's research. It has an internal reference system that is structured on author/compound/subject indexes to facilitate access to the information available in the work.

6.053

Jordan, Robert B. *Reaction mechanisms of inorganic and organometallic systems*. 2nd ed. NY: Oxford

University Press, 1998. 371 p. (Topics in inorganic chemistry) 0195115554 QD502 541.3 97-31222

Includes: Tools of the trade; Rate law and mechanism; Ligand substitution reactions; Stereochemical change; Reaction mechanisms of organometallic systems; Oxidation-reduction reactions; Inorganic photochemistry; Bioinorganic systems. Includes references on recent review articles.

6.054

Lappin, Graham. *Redox mechanisms in inorganic chemistry*. NY: Ellis Horwood, 1994. 285 p. (Ellis Horwood series in inorganic chemistry) 0137707517; 0137700172 pbk QD63.O9 541.3 93-50189

Descriptive rather than theoretical approach to inorganic redox chemistry, with a variety of examples. Thoroughly discusses inner-sphere and outer-sphere charge transfer reactions. The final two chapters cover more advanced topics, mainly intramolecular electron transfer in mixed valence compounds and reactions that involve multiple electron transfer. A short problem set is included with each chapter. Copious literature references and extensive tables of kinetic and thermodynamic data.

SPECIAL TOPICS

Spectroscopy

6.055

The infrared spectra handbook of inorganic compounds. Philadelphia, PA: Sadtler, 1984. 345, 19, 7 p. 0845601121 QC457 544 84-52117

Spectra of 345 compounds of 32 elements are included, together with Chemical Abstracts name, synonyms, density, melting point, solubility, color and/or crystalline structure, CASRN, RTECS number, molecular formula and weight, source of material, and method of sample preparation.

6.056

Inorganic mass spectrometry: fundamentals and applications, ed. by Christopher M. Barshick, Douglas C. Duckworth, and David H. Smith. NY: Marcel Dekker, 2000. 512 p. (Practical spectroscopy, 23) 0824702433 QD95 543 99-87830

Treats mass spectrometry techniques, including thermal ionization, glow discharge, inductively coupled plasma, secondary ion, and isotope dilution mass spectrometry. Discusses emission of ions from high-temperature condensed phase materials; analysis of nonconductive sample types; ion traps, elemental speciation, geological applications, time-of-flight mass spectrometry, and multiple-collector inductively coupled plasma mass spectrometry.

6.057

Nakamoto, Kazuo. *Infrared and Raman spectra of inorganic and coordination compounds*, 5th ed. NY: Wiley, 1997. 2 v. (A Wiley-Interscience publication) 0471194069 (set); 0471163945 (pt. A); 0471163929 (pt. B) QD96 543 96-33456

Pt. A: Theory and applications in inorganic chemistry; Pt. B: Applications in coordination, organometallic, and bioinorganic chemistry. This new edition (4th ed., 1986) includes new work on lattice vibrations, large carbon clusters, and ceramic superconductors. Many new spectra have been added. Part A has 10 appendixes with point groups and character tables, matrix algebra and applications, group frequency charts, correlation tables, and space groups.

6.058

Nyquist, Richard A., Curtis L. Putzig, and M. Anne Leugers. *The handbook of infrared and Raman spectra of inorganic compounds and organic salts*. San Diego: Academic Press, 1997. 4 v. 0125234449 (set) QC457 543 96-22175

V. 1-3 have spine title: Spectral Atlas. V. 4 originally published in 1971 as Nyquist, Richard A. and Ronald O. Kagel, *Infrared Spectra of Inorganic Compounds (3800-45cm-1)*; v. 1: Infrared and Raman Spectral Atlas of Inorganic Compounds and Organic Salts. Text and explanations; v. 2: ... Raman spectra: v. 3: ... Infrared spectra. Lists infrared and Raman spectra of a very large number of inorganic compounds and organic salts, including some nonionic inorganic compounds; compendium of Raman spectra otherwise unavailable. V. 1 discusses experimental and theoretical aspects as well as vibrational assignments. Many diagrams show vibrational modes for numerous compounds. Valuable tables, correlation diagrams, and references. V. 2 and 3 contain Raman spectra and IR spectra of some 500 compounds, respectively. Volume 4 is a revision of Nyquist's original work, *Infrared Spectra of Inorganic Compounds ...* (1971), with another 892 compounds and miscellaneous minerals. Extensive index cross-references Raman and IR spectra. IR spectra in v. 3

are prepared as potassium bromide pellets; those in v. 4 are done as split mulls.

6.059

Parish, R.V. *NMR, NQR, EPR, and Mössbauer spectroscopy in inorganic chemistry.* NY: E. Horwood, 1990. 223 p. (Ellis Horwood series in inorganic chemistry) 0136255183 QD95 543 90-4873

Introduces inorganic spectroscopy techniques, and is particularly thorough in the NMR chapter, discussing 2-D, correlated spectroscopy (COSY), and internuclear double resonance (INDOR) techniques.

Solid State Chemistry

6.060

Aldersey-Williams, Hugh. *The most beautiful molecule: the discovery of the buckyball.* NY: John Wiley, 1995. 340 p. 047110938x QD181.C1 546 95-12422

Excellent, well-written account of ideas and people behind the discovery of buckminsterfullerene that details process and personal interactions behind the 1985 discovery of this third form of carbon. Entire chapters describe important background information; e.g., a chapter on R. Buckminster Fuller provides a brief biography of this innovative thinker for whom the molecule is named. Other chapters offer concise but complete accounts of different types of spectroscopy used to identify C_{60} as unique and novel, and discuss symmetry and mathematical properties of the truncated icosahedron (the C_{60} soccer-ball structure).

6.061

Baggott, Jim E. *Perfect symmetry: the accidental discovery of buckminsterfullerene.* Oxford; NY: Oxford University Press, 1994. 315 p. 0198557906; 0198557892 pbk QD181.C1 546 94-14068

The discovery of C_{60} ("buckminsterfullerene") electrified the chemical world and even caught the imagination of the popular press. This totally new, unexpected, and outrageously novel form of carbon, spheroidally arranged carbon atoms, mimicking the geometry of a soccer ball. This book is a gripping history of the race to understand the nature of C_{60} and a review of the many directions that fullerene chemistry is now taking.

6.062

Dann, Sandra E. *Reactions and characterization of solids.* NY: Wiley-Interscience, 2002. 201 p. (Basic concepts in chemistry) 0471224812 QD478 541 2002-511022

Introduces crystal chemisty, preparation and characterization of solid-state materials, elementary thermodynamics, and bond strengths for predicting compound stability.

6.063

Elliott, S.R. *The physics and chemistry of solids.* Chichester, West Sussex, England; NY: J. Wiley, 2000, c1998. 770 p. Reprinted with corrections, January 2000. 047198194x; 0471981958 pbk QC176 530.41

Covers many classical and current topics from theoretical and practical standpoints. Eight chapters describe the synthesis, structures, defects, dynamics, properties, and applications of solid-state materials. Each chapter begins with a detailed table of contents and ends with an extraordinary set of problems (including essay questions) and a representative list of references to both the primary and secondary literature. Ten-page listing of abbreviations, acronyms, and symbols; separate chemical compound and subject indexes.

6.064

Fullerenes: chemistry, physics, and technology, ed. by Karl M. Kadish and Rodney S. Ruoff. NY: Wiley-Interscience, 2000. 968 p. 0471290890 QD181.C1 546 00-33402

Fullerenes ("buckyballs") are a new class of carbon-based compounds consisting of hollow spheres of 60 or more carbon atoms. Buckminsterfullerene, a C_{60} compound, was discovered during laser vaporization experiments. The fullerenes have potential for materials science applications such as semiconductors, and in pharmaceutics, polymer science, and the chemical industry.

6.065

Hannay, Norman B. *Treatise on solid state chemistry,* ed. by N.B. Hannay. NY: Plenum Press, 1973-76. 6 v. in 7. 0306350505 (set) QD478 541 73-13799

V. 1: The chemical structure of solids; v. 2: Defects in solids; v. 3: Crystalline and noncrystalline solids; v. 4: Reactivity of solids; v. 5: Changes of state; v. 6A: Surfaces I; v. 6B: Surfaces II. Experts describe the composition and atomic structure of the solid, and relate them to their chemical and physical properties. Among the many areas covered are chemical structure, disordered solids, defects, and chemical dynamics.

6.066

Hirsch, Andreas. *The chemistry of the fullerenes.* Stuttgart; NY: G. Thieme Verlag, 1994. 203 p. (Thieme organic chemistry monograph series) 3131368012 QD181.C1 546 94-28162

Comprehensive overview of the chemistry of "Buckyballs" and other new forms of carbon. Treats theory, preparation and isolation, and applications.

6.067

Ladd, Marcus Frederick Charles. *Structure and bonding in solid state chemistry.* Chichester, UK: Ellis Horwood; NY: Halsted Press, 1979. 326 p. (Ellis Horwood series in chemical science) 0470265973 (Halsted Press) QD478 541 78-41289

Treats ionic, covalent, van der Waals, and metallic bonding. Detailed preliminary section on units, symbols, and problem-solving techniques. Liberal use of stereo-structural diagrams.

6.068

Solid state chemistry: compounds, ed. by A.K. Cheetham and Peter Day. Oxford: Clarendon Press; NY: Oxford University Press, 1992. 304 p. (Oxford science publications) 0198551665 QD478 541 91-12173

6.069

Solid state chemistry: techniques, ed. by A.K. Cheetham and Peter Day. Oxford: Clarendon Press; NY: Oxford University Press, 1987. 398 p. (Oxford science publications) 0198551657 QD478 541 85-3121

These two books are a series of survey articles by experts, some on background material, and some on current topics. Electronic structure and bonding, superconductivity, ferroics, and catalysis are among the topics treated.

Organometallic Chemistry

6.070

Comprehensive organometallic chemistry: the synthesis, reactions, and structures of organometallic compounds, ed. by Sir Geoffrey Wilkinson; deputy editor, F. Gordon A. Stone; executive editor, Edward W. Abel. Oxford; NY: Pergamon Press, 1982. 9 v. 0080252699 (set) QD411 547 82-7595

Each of the first seven volumes contains signed, authoritative articles; tabular data; literature references. V. 8 offers special topics, and v. 9 contains subject, author, and formula indexes; structure indexes; index to review articles and books.

6.071

Comprehensive organometallic chemistry II: a review of the literature 1982-1994. Eds. in chief Edward W. Abel, F. Gordon A. Stone, and Geoffrey Wilkinson. Oxford; NY: Pergamon, 1995. 14 v. 0080406084 (set) QD411 547 95-7030

V. 1: Lithium, beryllium, and boron groups; v. 2: Silicon group, arsenic, antimony, and bismuth; v. 3: Copper and zinc groups; v. 4: Scandium, yttrium, lanthanides and actinides, and titanium group; v. 5: vanadium and chromium groups; v. 6: Manganese group; v. 7: Iron, ruthenium, and osmium; v. 8: Cobalt, rhodium, and iridium; v. 9: Nickel, palladium, and platinum; v. 10: Heteronuclear metal-metal bonds; v. 11: Main-group metal organometallics in organic synthesis; v. 12: Transition metal organometallics in organic synthesis; v. 13: Structure index; v. 14: Cumulative indexes.

6.072

Hill, Anthony F. *Organotransition metal chemistry.* NY: Wiley-Interscience; [Cambridge, England]: Royal Society of Chemistry, 2002. 185 p. (Basic concepts in chemistry) 0471281638 pbk QD411.8.T73

Treats compounds with bonds between carbon and a transition metal; discusses reactivity of the organometallic moiety and their structure and bonding characteristics.

6.073

Liebau, Friedrich. *Structural chemistry of silicates: structure, bonding, and classification.* Berlin; NY: Springer-Verlag, 1985. 03887137475 (US) QD181.S6 546 84-23532

Highlights silicates as a prevalent and important class of compounds. Detailed reference list. Chemical bonding material is very well written.

6.074

Synthetic methods of organometallic and inorganic chemistry: (Herrmann/Brauer) [ed. by W.A. Herrmann]. Stuttgart; NY: Georg Thieme Verlag; NY: Thieme Medical Publishers, 1996-2000. 10v. 3131030216 (GTV: v. 1); 3131030313 (GTV: v. 2); 3131030410 (GTV: v. 3); 3131030518 (GTV: v. 4);

3131030615 (GTV: v. 5); 3131030712 (GTV: v. 6); 313103081x (GTV: v. 7); 3131030917 (GTV: v. 8); 3131151412 (GTV: v. 9); 3131151617 (GTV: v. 10); 086577627x (TMP: v. 1); 0865776539 (TMP: v. 2); 0865776547 (TMP: v. 3); 08657761x (TMP: v. 4); 0865776628 (TMP: v. 5); 0865776636 (TMP: v. 6); 0865776644 (TMP: v. 7); 0865776652 (TMP: v. 8); 0865779716 (TMP: v. 9) ; 1588909406 (TMP: v. 10); QD411 541.3 95-49908

V. 1: Literature, laboratory techniques, and common starting materials; v. 2: Groups 1, 2, 13, and 14; v. 3: Phosphorus, arsenic, antimony, and bismuth; v. 4: Sulfur, selenium, and tellurium; v. 5: Copper, silver, gold, zinc, cadmium, and mercury; v. 6: Lanthanides and actinides; v. 7: Transition metals, part 1; v. 8: Transition metals, part 2; v. 9: Transition metals, part 3; v. 10: Catalysis.

6.075

Yamamoto, Akio. *Organotransition metal chemistry: fundamental concepts and applications*. NY: Wiley, 1986. 455 p. (A Wiley-Interscience publication) 0471891711 QD411 547 85-26349

Rev. translation of *Yūki kinzoku kagaku*. Provides basic concepts in catalysis and syntheses using transition metal reagents. Covers basic principles of coordination chemistry, organometallic compounds of transition and nontransition metals, reactions, industrial applications, synthetic uses, manipulation of air-sensitive materials, and overviews of related topics.

Lanthanides, Actinides, and Nuclear Chemistry

6.076

Adloff, J.P., and Robert Guillaumont. *Fundamentals of radiochemistry*. Boca Raton, FL: CRC Press, 1993. 414 p. 0849342449 QD601.2 541.3 92-299544

Discusses the early history of radioactive elements up through positron emission tomography (PET) and chemistry with few atoms; includes kinetics, statistical methods, analytical techniques, and calculations. Focuses on trace-level chemistry, geochemically or in nuclear wastes, and includes transport, partition methods, and the chemistry of actinides and postactinide elements.

6.077

The chemistry of the actinide elements, ed. by Joseph J. Katz, Glenn T. Seaborg, and Lester R. Morss. 2nd ed. London; NY: Chapman and Hall, 1986. 2 v. 1,677 p. 0412105500 (v. 1); 0421173705 (v. 2) QD172 546 86-8284

Includes bibliographies and index. Covers the spectra and electronic structure of actinide ions in compounds and in solution, thermodynamic properties, the metallic state, structural chemistry, solution chemistry and kinetics of ionic reactions, organoactinide chemistry, properties of compounds having metal-carbon bonds only to p-bonded ligands, organoactinide chemistry, properties of compounds with actinide-carbon, actinide-transition-metal O bonds, future elements (including superheavy elements). V. 1: 886 p. Chemistry of actinium, thorium, protactinium, uranium, neptunium, and plutonium. V. 2: 912 p. Americium, curium, berkelium, californium, einsteinium, and transeinsteinium.

6.078

Cotton, Simon. *Lanthanides and actinides*. NY: Oxford University Press, 1991. 192 p. 0195073665 QD172 546 91-17815

The book concentrates on the basics: occurrence, extraction, physical properties, and chemistry and radiochemistry of scandium; the lanthanide group metals; and the actinide group metals.

6.079

Handbook on the physics and chemistry of the actinides, ed. by A.J. Freeman and G.H. Lander. Amsterdam; NY: North-Holland; sole distributors for U.S.A. and Canada, Elsevier Science Pub. Co., 1984-91. 6 v. (v. 3-6 ed. by A.J. Freeman and C. Keller) 0444869034 (v. 1); 0444869077 (v. 2); 0444869263 (v. 3); 0444869832 (v. 4); 0444870563 (v. 5); 044487447x (v. 6) QD172 546 84-20698

The handbook's purpose is to describe in detail the present understanding of the actinides by comprehensive, critical, broad, and up-to-date reviews covering both the physics and chemistry of these elements.

6.080

Handbook on the physics and chemistry of rare earths, ed. by Karl A. Gschneidner Jr. and LeRoy Eyring. Amsterdam; NY: North-Holland; sole distributors for the U.S.A. and Canada, Elsevier North-Holland, 1978-<1997 > v. <1-24 > 0444850228 (set); 0444889663 (v. 15) QD172 546 78-12371

Separately published cumulative index volume for v. 1-15. Editors vary. V. 1: Metals; v. 2: Alloys and intermetallics; v.

3-4: Nonmetallic compounds; v. 5-9; v. 10: High energy spectroscopy, ed. by K.A. Gschneidner Jr., L. Eyring, and S. Hufner; v. 11: Two-hundred-year impact of rare earths on science; v. 12-16; v. 17: Lanthanides/actinides: physics I; v. 19: Lanthanides/actinides: physics II.

6.081

Hoffman, Darleane C., Albert Ghiorso, and Glenn T. Seaborg. *The transuranium people: the inside story.* London: Imperial College Press, 2000. 467 p. 1860940870 QD602.4 546.44092273

Includes history, experimental details, diagrams, laboratory organization information, data, and literature references. Naming of transuranium elements is discussed, and there are ample reminiscences and interesting stories surrounding the search for these elements.

6.082

Lieser, Karl Heinrich. *Nuclear and radiochemistry: fundamentals and applications.* 2nd, rev. ed. Berlin; NY: Wiley-VCH, 2001. 462 p. 3527303170 QD601.3 541.3 2001-272382

Includes chart of the nuclides in a pocket. Treats radioanalysis, radiodating, radiotracers, nuclear energy, industrial applications, and geochemistry and cosmochemistry. Fundamental topics discussed include radioactive decay, radiation detection and measurement, nuclear reactions, and radiation dosimetry and protection. Elementary particles are treated, as is cosmic radiation.

6.083

Transuranium elements: a half century, ed. by L.R. Morss and J. Fuger. Washington, DC: American Chemical Society, 1992. 562 p. 0841222193 QD172.T7 546 92-7475

"Developed from a symposium sponsored by the Divisions of Nuclear Chemistry and Technology, the History of Chemistry, and Inorganic Chemistry of the ACS." Describes the fascinating synthesis and study of synthetic elements, their history, chemistry, separation, thermodynamics, physics, and analysis.

6.084

Trenn, Thaddeus. *Transmutation, natural and artificial.* London; Philadelphia: Heyden, [1981]. 128 p. (Nobel prize topics in chemistry) 085501685x; 0855016868 pbk QC794.5 539.7 82-106284

Traces the pioneering desires of alchemists to change base metals into gold to recent efforts to change elements, up through the 20th century. Includes the three research articles by Rutherford and Soddy in 1903, by Curie and Joliet-Curie in 1934, and by Hahn and Strassmann in 1938, the seminal papers that precipitated the development of new research.

Chemistry of Specific Elements

6.085

Chemistry of hypervalent compounds, ed. by Kin-ya Akiba. NY: Wiley-VCH, 1999. 414 p. 0471240192 QD255.4 541.2 98-6578

Overviews research on hypervalent compounds of representative elements. "Hypervalent" refers to the number of valence electrons surrounding the central atom when bonded to other molecules (or ligands) in these compounds, or an expansion of the octet of electrons for these elements. Chapter references are current through 1997. The chapters by experts cover compounds of many of the nonmetallic elements; Two chapters are devoted to the structure, reactivity, and synthetic applications of silicon and iodine compounds.

6.086

Chemistry of iron, ed. by J. Silver. London; NY: Blackie Academic & Professional, 1993. 306 p. 0751400629 QD181 546 93-116922

Provides the important chemistry of iron in both its elemental and combined forms. Most of the book is devoted to reviews of research in inorganic and organometallic chemistry and spectroscopic methods for the analysis of iron compounds. Also discussed are the industrial uses of iron and its compounds, and the chemistry underpinning diverse biochemical and pharmaceutical applications of iron. Chapters: Introduction to iron chemistry; Industrial chemistry of iron and its compounds; Inorganic chemistry of iron; Organo-iron compounds; Spectroscopic methods for the study of iron chemistry; Biological iron; Models for iron biomolecules; Iron chelators of clinical significance.

6.087

Emsley, John. *The 13th element: the sordid tale of murder, fire, and phosphorus.* NY: John Wiley & Sons, 2000. 327 p. 0471394556 QD181.P1 546 00-703013

Phosphorus is one of the deadliest and longest-known (300 years) elements; known to alchemists and prescribed by apothecaries through the ages, it forms deadly compounds with unusual properties, such as phosphorescence.

6.088

Fluorine chemistry at the millennium: fascinated by fluorine, ed. by R.E. Banks. Amsterdam; NY: Elsevier, 2000. 643 p. 0080434053 QD181.F1 546 00-26273

Contributions by leading researchers treat a wide range of topics. Moissan discovered fluorine in 1886, and this volume celebrates 100 years of fluorine chemistry and discoveries.

6.089

Gold: progress in chemistry, biochemistry, and technology, ed. by Hubert Schmidbaur. Chichester; NY: Wiley, 1999. 894 p. 0471973696 QD181.A9 546 98-7028

In three sections—a long one on gold technology, a short one on biochemistry, and a major one on chemistry—this multiauthored and authoritative work completely covers gold in various disciplines of science and technology. Generously illustrated; each chapter fully referenced with citations current to the mid-'90s. "Gold Technology" discusses history and progress; gold for coinage; history, economics, and geology; recovery from ores and environmental aspects; gold metal and alloys in jewelry; refining of gold and recycling of electronics. "Chemistry" treats gold compounds of nitrogen, phosphorus, and heavy Group V elements, and molecular compounds of gold with main group and transition metals.

6.090

Gutmann, Viktor. *Halogen chemistry*, ed. by Viktor Gutmann. London; NY: Academic Press, 1967. 3 v. QD165 546 66-30147

A collection of reviews on all aspects of inorganic and organometallic chemistry of halogen-containing compounds.

6.091

Rao, C.N.R., and B. Raveau. *Transition metal oxides.* NY: VCH, 1995. 338 p. 1560816473 QD172.T6 546 95-10717

Thoroughly treats substances containing oxygen and a transition metal; many of the compounds contain one to three other elements as well. The book is organized in three sections: structure, properties and phenomena, and preparation of materials. Provides very clear sketches of structures and uses electron-microscope images very nicely to illustrate crystal defects. The properties section includes a

stimulating range of topics, including brief discussion of band structure, magnetic properties, superconductivity, catalysts, and nanomaterials, among others. Methods of preparing materials are discussed succinctly.

6.092

Rigden, John S. *Hydrogen: the essential element.* Cambridge, MA: Harvard University Press, 2002. 280 p. 0674007387 QD181.H1 546 2001-51708

Beginning with William Prout's hypothesis defining hydrogen as a basic building block of nature, Rigden relates various episodes in which the simplest element has yielded its secrets to persistent researchers.

6.093

Sawyer, Donald T. *Oxygen chemistry*; foreword by R.J.P. Williams. NY: Oxford University Press, 1991. 223 p. (The International series of monographs on chemistry, 26) 0195057988 QD181.O1 546 91-6566

Surveys reactions of diatomic oxygen, and includes electrochemical and thermochemical parameters, reactivity, and biological applications. Treats the chemistry of diatomic oxygen, peroxides, superoxides, and related compounds.

Analytical Methods and Procedures, Including Chromatography

6.094

Albert, Adrien and E.P. Serjeant. *The determination of ionization constants: a laboratory manual.* 3rd ed. London; NY: Chapman and Hall, 1984. 218 p. 0412242907 QD561 541.3 83-7497

A successor to the original *Ionization Constants of Acids and Bases* (1962). Limited to acid-base equilibria; no stability constants for dissociation complexes. Potentiometric and spectrometric techniques are presented in detail.

6.095

Feigl, F., and Vinzenz Anger. *Spot tests in inorganic analysis.* Tr. from German by Ralph E. Oesper. 6th English ed., rev. and enl. Amsterdam; NY: Elsevier, 1972. 669 p. 0444409297 QD81 544 76-135494

Small-scale testing procedures are featured. Includes tests for metals, cations, and anions of metalloacids; tests for acid radicals, anions; tests for free elements; systematic analysis of mixtures by spot reactions; spot reactions in tests of purity, technical materials, and minerals.

6.096
Inorganics. Mohsin Qureshi, ed. Boca Raton, FL: CRC Press, 1987- . v. <1 >. (CRC handbook of chromatography). 0849330491 (v. 1) QD117 543 85-22418

V. 1: Principles, techniques, and quantitative determinations; Detection techniques; Liquid chromatography; Gas chromatography; Paper chromatography; Thin-layer chromatography; Liquid chromatography; Gas chromatography; Directory.

6.097
Lederer, Michael. *Chromatography for inorganic chemistry*. Chichester; NY: J. Wiley, 1994. 221 p. 0471942855; 0471942863 pbk QD79 543 93-31657

Treats history; solvent extraction and the history of partition chromatography; paper chromatography and thin-layer chromatography; electrophoresis; gel filtration; ion exchange; HPLC; ion chromatography; gas chromatography; separation of isotopes; separation of optical isomers; some elements and their chromatography and electrophoresis.

6.098
Vogel, Arthur Israel. *Vogel's textbook of quantitative inorganic analysis*, 6th ed. rev. by J. Mendham ... [et al.]. Harlow, Essex, UK: Prentice Hall, 2000. 836 p. 0582226287 QD101 545 99-X4953

Includes many organic applications and experiments. Treats fundamentals; errors, statistics, sampling; separation techniques, titrimetry and gravimetry, and electroanalytical and spectroanalytical methods.

Chapter 7

Environmental Chemistry

This chapter begins the parts of this book that are actually applied chemistry; they will treat the literature of non-theoretical chemistry rather than that of pure laboratory practice. In this series will be included biochemistry, industrial chemistry, and polymer chemistry—smaller subareas of the "big four": analytical, inorganic, organic, and physical chemistry. The applied areas are also not "pure" or single-subject areas; each treats elements of all the "big four" within it, and could be thought of as truly interdisciplinary areas. These applied/ interdisciplinary areas of chemistry have enjoyed a wide acceptance and popularity during the last twenty or so years; the environmental movement with all its emphasis on health and safety for humans and their surroundings has spawned a chemistry all its own. There are books and references on health and safety aspects of huge numbers of chemicals used in education and industry. There are works on toxicology, pollution, and waste management. Literally, there are books for just about any aspect of the environment and its relationship with chemistry. The latter part of this chapter covers more-specialized sources of information, digging deeper into more-advanced aspects of this area. Materials discussed in this section include many that typical undergraduate students may not use, but graduate students, their faculty, and chemists in industry will need to know about these sources. As is often the case with compilations like this book, there will of necessity be many books and other materials omitted, but this fact does not diminish their importance.

The Library of Congress areas touched on in environmental chemistry, besides the usual QD areas of traditional chemistry, are classified under RA; public aspects of medicine, including public health and toxicology are examples. Environmental health subjects are here, classified at RA 421-790.85. Other areas of the Library of Congress Classification represented here include RS, pharmacy and materia medica, and the relatively new area of this classification, GE, environmental sciences, added in the '90s.

GUIDES TO THE LITERATURE

7.001
Alston, Patricia Gayle. "Environment online: the greening of databases. Part 2. Scientific and technical databases." *Database* **14**(5), 34-52 (October 1991)

Discusses the scientific and technical databases available for searching environmental data and issues.

7.002
Freeman, Robert R., and Mona F. Smith. "Environmental information", Chapter 8, in *Annual review of information science and technology* (ARIST), Vol. 21. Knowledge Industry Publications, Inc., for American Society for Information Science, 1986.

This chapter compiles information current at its writing (1985) on sources of environmental information. Chemistry is only a part of this area. This material serves as a good historical summary on the subject.

7.003
Kokoszka, Leopold C. "Guide to federal environmental databases." *Pollution Engineering,* p. 83-88, 90, 92 (Feb. 1, 1992)

Federal databases discussed include those covering air, contractor, emergency response; facilities; hazardous waste/solid waste; laboratory quality assurance/quality control; physical/chemical properties; Superfund/CERCLA; toxic substances/TSCA; toxicity, treatment, and water.

7.004
Pantry, S. "Health and safety," Chapter 13 in *Information sources in chemistry*, ed. by R.T. Bottle and J.F.B. Rowland. NY: Bowker-Saur, 1992. pp. 231-249.

Discusses chemical and toxicological information from printed reference sources as well as Health and Safety Executive publications and US NIOSH publications; online and CD-ROM services, and a number of health and safety abstracting and indexing services as well as materials safety data sheets suppliers. Legal information, translation services, journals, software, and pertinent organizations are also discussed.

7.005

Toxic and hazardous materials: a sourcebook and guide to information sources, ed. by James K. Webster. NY: Greenwood Press, 1987. 431 p. (Bibliographies and indexes in science and technology; 0888-7551; v. 2) 0313245754 Z7914 016.3631 86-25710

Some 1,600 entries include books, periodicals, databases, audiovisual materials, organizations, and so forth.

7.006

Using the agricultural, environmental, and food litera-ture, ed. by Barbara S. Hutchinson and Antoinette Paris Greider. NY: Marcel Dekker, 2002. 533 p. (Books in library and infor-mation science, 61) 0824708008 S494.5.A39 630 2002-73403

Explains how to find information in encyclopedias, handbooks, manuals, atlases, abstracting and indexing services, bibliographies, periodicals, monographs, theses, conference proceedings, patents, trade literature, and government documents. Subject areas discussed include, among others, environmental sciences and natural resources, as well as soil science.

7.007

Wolman, Yecheskel. "Environmental impact and control, oc-cupational hygiene, safety, toxicity", Chapter 11 in *Chemical information: a practical guide to utilization.* 2nd rev. and enl. ed. NY: Wiley, 1988. 291 p. (A Wiley-Interscience publication) 0471917044 QD8.5 540 87-23119

Discusses the safety literature, primary and secondary; general safety practice; safety in the chemistry laboratory and conducting a safety-related literature search (to obtain information on dangerous physical and biological characteristics of chemical compounds), with examples of where information is found and what it means.

ABSTRACTING AND INDEXING SERVICES

7.008

Pollution abstracts. Bethesda, MD: Cambridge Scientific Abstracts, v. 1, May 1970- . Bimonthly. ISSN 0032-3624 TD172 628 75-17056

V. 1-2, 1970-71, 1 v.; v. 3, 1972. Indexes and abstracts the worldwide literature on air, marine and freshwater, noise, land, and environmental pollution; sewage and wastewater treatment, toxicology and health, and other related topics. Microfiche; online and CD-ROM versions are also available.

7.009

Pollution abstracts (CD-ROM). Baltimore, MD: National Information Services Corp. (NISC), 1970- . Quarterly. ISSN 1093-3409 333 97-3784

"NISC disc". Electronic version of printed abstracting service. Online version available through Dialog, ESA-IRS, Data-Star, and STN.

7.010

TOXNET. National Library of Medicine.   http://sis.nlm.nih. gov/sis1/

A cluster of Internet databases including Hazardous Substances Data Bank (HSDB), Integrated Risk Information System (IRIS), Chemical Carcinogenesis Research Information System (CCRIS), GENE-TOX, TOXLINE, Environmental Mutagen Information Center (EMIC), Developmental and Reproductive Toxicology and Environmental Teratology Information Center (DART/ETIC), Toxic Release Inventory (TRI), ChemID*plus*, and other related resources.

GENERAL ENVIRONMENTAL CHEMISTRY

7.011

Basic concepts of environmental chemistry; Des W. Connell ... [et al.]. Boca Raton, FL: Lewis Publishers, 1997. 506 p. 0873719980 TD193 628.5 97-7698

Offers introductions to the principles of toxicology, global geochemistry, and management of hazardous substances. Chapters treat specific environmental contaminants, such as petroleum, polychlorinated biphenyls, dioxins, polymers, pesticides, and others. Emphasis is on organics, although there is material on organometallics. Treatment is well balanced, though the number of topics discussed precludes any in-depth coverage. Basic diagrams, equations, and graphs; chapter summaries, lists of useful books, and practice problems with solutions.

7.012

Chemicals in the human food chain, ed. by Carl K.

Winter, James N. Seiber, and Carole Frank Nuckton. NY: Van Nostrand Reinhold, 1990. 276 p. 0442004214 TX531 363.1 89-70670

Catalogs, documents, and evaluates sources of chemicals in the human food chain that have demonstrated a potential for exposures above the toxic threshold, or which could have such a potential. Identifies chemicals whose benefits are greater than their potential harm.

7.013

Fergusson, Jack E. *The heavy elements: chemistry, environmental impact, and health effects.* Oxford; NY: Pergamon Press, 1990. 614 p. 0080348602; 0080402755 pbk TD196 363.17 90-6783

Contents: introduction and history; chemistry; heavy elements in the environment; heavy elements in human beings.

7.014

Introductory chemistry for the environmental sciences, by Roy M. Harrison ... [et al.]. Cambridge; NY: Cambridge University Press, 1991. 354 p. (Cambridge environmental chemistry series) 0521256739; 052127639x pbk QD31.2 540 90-2370

Overviews atomic and molecular structure, physical chemistry, inorganic chemistry, some organic chemistry, and analytical chemistry, in the area of environmental sciences.

7.015

Manahan, Stanley E. *Environmental chemistry.* 6th ed. Boca Raton, FL: Lewis Publishers, 1994. 811 p. 1566700884 QD31.2 628.5 94-18437

Discusses aquatic chemistry, the geosphere and hazardous wastes, toxicological chemistry, and resources and energy.

7.016

Perspectives in environmental chemistry, ed. by Donald L. Macalady. NY: Oxford University Press, 1998. 512 p. (Topics in environmental chemistry) 0195102088; 0195102096 pbk TD193 628.5 97-4123

Covers most areas of current interest, including natural water chemistry (colloids, trace metals, and organics), tropospheric oxidants and aerosols, stratospheric chemistry (aerosols, the role of carbon dioxide, and ozone), and applications of environmental chemistry. The writing is clear and readable. Extensive chapter bibliographies.

7.017

Pollution: causes, effects and control, ed. by Roy M. Harrison. 4th ed. Cambridge: Royal Society of Chemistry, 2001. 579 p. 0854046216 TD174 363.73

Previous edition: 1996. Chapters discuss priority pollutants, marine pollution, drinking water quality, water pollution biology, sewage and sewage sludge treatment, toxic wastes, air pollution, chemistry and climate change in the troposphere, chemistry and pollution of the stratosphere, atmospheric dispersal of pollutants, modeling of air pollution, health effects of air pollution, gaseous pollutants' effects on the biosphere, road traffic emissions, soil pollution and land contamination, solid wastes, clean technologies and industrial ecology, persistent organic pollutants, radioactivity, health effects, regulations and control.

7.018

Sawyer, Clair N., Perry L. McCarty, and Gene F. Parkin. *Chemistry for environmental engineering.* 4th ed. NY: McGraw-Hill, 1994. 658 p. (McGraw-Hill series in water resources and environmental engineering) 0070549788 TD193 628 94-261

Previously published as *Chemistry for Sanitary Engineers*, 2nd ed., 1967.

7.019

Schwarzenbach, René P., Philip M. Gschwend, and Dieter M. Imboden. *Environmental organic chemistry.* 2nd ed. Hoboken, NJ: Wiley, 2003. 1,313 p. (A Wiley-Interscience publication) 0471350532 TD196.O73 628.1 2002-513118

Analyzes and explains reactions and transformations of organic chemicals in the environment. Treats vapor pressure, aqueous solubility, activity coefficient in water, air-water partitioning and Henry's Law constant, organic solvent-water partitioning, octanol-water partition constant, organic acids and bases, diffusion, gas-liquid interface, air-water exchange, sorption, photochemical reactions, biological transformations, and modeling concepts.

DICTIONARIES AND ENCYCLOPEDIA

7.020

Ayres, David C., and Desmond Hellier. *Dictionary of environmentally important chemicals.* London; NY: Blackie

Academic & Professional, 1998. 332 p. 0751402567
TD196.C45 363.17 97-74237

Chemicals listed are mentioned by at least three of five international regulatory bodies. Entries remain for products discontinued in all but Third World nations. Each alphabetical entry contains molecular formula, structural diagrams, CASRN, physical properties, production and distribution, uses, toxicity, occupational exposure limits, and further reading. Some information on pollution incidents, storage risks, human exposure sources, metabolism, and carcinogenic or mutagenic activity are provided.

7.021

Coleman, Ronny J. *Hazardous materials dictionary*. 2nd ed. Lancaster, PA: Technomic Publ. Co., 1994. 209 p. 1566761603 T55.3 363.17 94-60735

This resource contains more than 2,600 concise definitions of words, phrases, abbreviations, and acronyms pertinent to hazardous materials. Classification, handling/storage/disposal, regulations, and emergency response are treated.

7.022

Dictionary of substances and their effects, ed. by S. Gangolli. 2nd ed. Cambridge, UK: Royal Society of Chemistry, 1999. 7 v. 0854048030 (set) RA1193 363.179103

This collection contains descriptions for about 4,100 chemicals and their impact on the environment. Toxicity, physical properties, and regulatory requirements are included. Available via the Internet from knovel.com.

7.023

Dictionary of toxicology, ed. by Ernest Hodgson, Richard B. Mailman, and Janice E. Chambers; asst. ed., Robert E. Dow. 2nd ed. NY: Van Nostrand Reinhold, 1988. 504 p. 1561592161 RA1193 615.9003 98-68580

A very useful dictionary with terminology from anatomy, pathology, and physiology. Revised ed. of *Macmillan Dictionary of Toxicology*, ed. by Ernest Hodgson, Richard B. Mailman, and Janice E. Chambers, 1988.

7.024

Encyclopedia of environmental pollution and cleanup; ed. in chief, Robert A. Meyers; Diane Kender Dittrick, ed. NY: Wiley, 1999. 2 v., 1,890 p. 0471316121 TD173 628.5 99-17884

This condensation of *Encyclopedia of Environmental Analysis and Remediation*, ed. by Robert A. Meyers (8 v., 1998), contains 230 of the original 250 articles but condenses many, making them more descriptive and less analytical. Well-designed tables, graphs, and diagrams; clear and well-organized writing. Extensive, current, and scholarly bibliographies.

7.025

Lewis, Robert Alan. *Lewis' dictionary of toxicology*. Boca Raton, FL: Lewis Publishers, 1998. 1,127 p. 1566702232 RA1193 615.9 96-35759

Covers both toxicology and toxins; mentions many organisms that produce toxins; toxicology terminology is well covered.

7.026

Soil and environmental science dictionary, ed. by E.G. Gregorich ... [et al.]. Boca Raton, FL: CRC Press, 2001. 577 p. 0849331153 S592 631.4 2001-25292

Covers a broad range of disciplines, including physical geography, geology, and meteorology. Like other Canadian reference works, it is published in English and French. The French equivalents to the terms are given, although the entries are exclusively in English. An extensive French-English index of terms is also included. Useful appendixes include information on soil classification and properties, the geologic timescale, and a map of the ecosystems of Canada.

HANDBOOKS AND DESK REFERENCES

7.027

Bretherick, L. *Bretherick's handbook of reactive chemical hazards: an indexed guide to published data*, ed. by P.G. Urben; compilers, M.J. Pitt, L.A. Battle. 5th ed. Oxford; Boston: Butterworth-Heinemann, 1995. 2 v. 0750615575 T55.3 660 95-199154

A previous edition (3rd) under title *Handbook of Reactive Chemical Hazards*. Contains about 9,000 elements and compounds, arranged by structural or reactivity similarity in the order of empirical formulae. Each class of chemical entity is described separately. Entries contain IUPAC-based name, CASRN, linear structural formula, original literature reference, and safety-related hazard information and data on use and handling (20th anniversary edition).

Environmental Chemistry

7.028

Chemical information manual. 3rd ed. Rockville, MD: Government Institutes, 1995. 341 p. 0865874697 RA1229.3 615.9 95-77857

Broad range of in-depth data on more than 1,200 chemicals includes proper chemical identification, OSHA exposure limits, description and physical properties, carcinogenic status, health effects and toxicology, and sampling and analysis. Publication is derived from OSHA information and documentation.

7.029

Handbook of ecotoxicology; David J. Hoffman, Barnett A. Rattner, G. Allen Burton Jr., and John Cairns Jr. 2nd ed. Boca Raton, FL: Lewis Publishers, 2002. 1,344 p. 1566705460 RA1226 615.9 2002-75228

Focuses on toxic materials and how they affect ecosystems. Includes methods of quantifying and measuring effects in the field and the lab, as well as methods for estimating, predicting, and modeling in ecotoxicology studies.

7.030

Handbook of environmental fate and exposure data for organic chemicals, ed. by Philip H. Howard. Chelsea, MI: Lewis Publishers, 1989-<1999 >. v. <1-5 >. 0873711513 (v. 1); 0873712048 (v. 2) TD176.4 363.7 89-2436

V. 1: Large production and priority pollutants; v.2: Solvents; v. 3: Pesticides; v. 4: Solvents; a projected seven-volume set. Chemicals are organized and analyzed by synonym, structure, CASRN, molecular formula, and Wiswesser line notation; chemical and physical properties; and environmental fate and exposure potential.

7.031

Lewis, Richard J., Sr. *Hazardous chemicals desk reference.* 5th ed. NY: Van Nostrand Reinhold, 2002. 1,695 p. 0471441651 T55.3.H3 604.7 2001-46947

An abridged version of *Sax's Dangerous Properties of Industrial Materials*. Provides hazard references to more than 5,000 chemical compounds used in industry, manufacturing, and laboratories. Included for each chemical are toxic and hazard review summaries, hazard ratings, CASRNs, NIOSH and DOT numbers, synonyms, physical properties, and other data. New or updated safety profiles are provided for each chemical, rated as poison, irritant, corrosive, explosive, or carcinogen.

7.032

Lewis, Richard J., Sr. *Sax's dangerous properties of industrial materials.* 10th ed. NY: Wiley, 2000. 3 v. (Wiley-Interscience publication) 0471354074 (set) T55.H3 604.7 99-39820

Revises more than two-thirds of the 9th ed. (1998) list of 23,500 entries, with easy-to-use visual aids inside front and back covers and a concise table of contents. The first volume provides a key to abbreviations used, an introduction to the material provided, indexes to CASRN and synonyms, and a list of CODEN reference codes (used in place of journal titles). The second and third volumes contain the entries themselves, arranged alphabetically by entry number, and including entry name, CASRN, hazard rating, DOT number, physical properties, synonyms, toxicity data with references, consensus reports (supplying additional information), standards and recommendations, and safety profiles summarizing toxicity and other dangerous properties of the material.

7.033

Patnaik, Pradyot. *A comprehensive guide to the hazardous properties of chemical substances.* 2nd ed. NY: John Wiley & Sons, 1999. 984 p. 0471291757 RA12311 615.9 98-39972

Contents: Physical properties of compounds and hazardous characteristics; Toxic properties of chemicals; Target organs and toxicology; Cancer-causing chemicals; Teratogenic substances; Habit-forming addictive substances; Flammable and combustible properties of chemical substances; Explosive characteristics of chemical substances; Peroxide-forming substances; Carboxylic acids; Mineral acids; Peroxy acids; Alcohols; Aldehydes; Alkalies; Alkaloids; Amines; Asbestos; Azo Dyes; Chlorohydrins; Cyanides, Nitriles; Dioxins; Epoxy compounds; Esters; Ethers; Gases; Glycol ethers; Haloethers; Halogenated hydrocarbons; Halogens, halogen oxides, and interhalogen compounds; Heterocyclics; Hydrocarbons; Industrial solvents; Isocyanates; Ketones; Metal acetylides and fulminates; Metal alkoxides; Metal alkyls; Metal azides; Metal carbonyls; Metal hydrides; Reactive and toxic metals; Mustard gas and sulfur mustards; Nerve gases; Nitro explosives; Oxidizers; Particulates; Organic peroxides; Pesticides and herbicides; Urea; Phenols; Phosphorus and compounds; Polychlorinated biphenyls; Radon and radioactive substances; Sulfate esters; Organosulfur compounds; Miscellaneous substances. Appendix with federal regulations, IARC list of carcinogens, and NTP list of carcinogens.

The Literature of Chemistry

7.034

Sax's dangerous properties of industrial materials [computer file]. Ed. by Richard J. Lewis, Sr. [ver. 1.3] NY: Van Nostrand Reinhold, 1996- . 1 CD-ROM (Comprehensive chemical contaminants series) T55.3.H3 604 97-48256

Compiles more than 20,000 industrial and laboratory materials; updates more than 14,000 entries and adds 1,500 new chemicals. Synonym index is extensive and includes foreign and trade names. The first volume provides a key to abbreviations used, an introduction to the material provided, indexes to CASRN and synonyms, and a list of CODEN reference codes (used in place of journal titles). The second and third volumes contain the entries themselves, arranged alphabetically by entry number, and including entry name, CASRN, hazard rating, DOT number, physical properties, synonyms, toxicity data with references, consensus reports (supplying additional information), standards and recommendations, and safety profiles summarizing toxicity and other dangerous properties of the material.

7.035

Solvents safety handbook, ed. by D.J. DeRenzo. Park Ridge, NJ: Noyes Data Corp., 1986. 696 p. 0815510748 TP247.5 660.2 86-5208

Contains data on 335 hazardous industrial solvents. Information for each solvent includes health hazards and toxicity; fire, exposure, and water pollution; protective equipment; handling, labeling, and shipping; saturated liquid density; liquid heat capacity; thermal conductivity; viscosity; and more.

7.036

Verschueren, Karel. *Handbook of environmental data on organic chemicals.* 4th ed. NY: Wiley, 2001- . v. <2 > (A Wiley-Interscience publication) 0471374903 TD196.O73 363.738 00-69254

First group of pages are devoted to arrangement of categories and chemicals; order of explanatory elements: properties, air pollution factors, water and soil pollution factors, biological effects. Contains glossary, abbreviations list, environmental data, bibliographies, molecular formula cross-references, and CASRN cross-index.

CHEMICAL ANALYSIS OF ENVIRONMENTAL SAMPLES

7.037

Soil sampling and methods of analysis, ed. by Martin R. Carter. Boca Raton, FL: Lewis Publishers, 1993. 823 p. 0873718615 S593 631.4 92-38583

Covers a wide assortment of methods for analyzing soil chemical, biological, biochemical, and physical properties. Soil testing for plant nutrients, methods for characterizing organic soils and frozen soils, recent improvements in methodology, new methods, and best methods available are provided. Methods are referenced.

7.038

Standard methods for the examination of water and wastewater. Prepared and published jointly by American Public Health Association, American Water Works Association, Water Environment Federation. 20th ed. Joint editorial board, Lenore S. Clesceri, Arnold E. Greenberg, Andrew D. Eaton; managing editor, Mary Ann H. Franson. Washington, DC: APHA, 1998. Various pagings; irregular. ISSN 8755-3546 QD142 543.3 55-1979

Begun in 1960. Includes bibliographies and index. Supplement to the 17th ed., 1991, 144 pp. Includes EPA-approved analytical methods for organic contaminants, including sample collection and preservation procedures. General introduction; physical examination; determination of metals; determination of inorganic nonmetallic constituents; determination of organic constituents; automated laboratory analysis; bioassay methods for aquatic organisms; microbial examination of water; biological examination of water.

7.039

Standard methods for the examination of water and wastewater [computer file]. 20th ed. American Public Health Association; American Water Works Association; Water Environment Federation, 1999. 1 CD-ROM, Windows format. QD142

Includes graphs, charts, tables, quality control issues, safety issues, waste minimization and disposal, laboratory apparatus, washing and sterilization, among others. Arranged in parts: 1000: Introduction; 2000: Physical and aggregate properties; 3000: Metals; 4000: Inorganic nonmetallic constituents; 5000: aggregate organic constituents; 6000: Individual organic compounds; 7000: Radioactivity; 8000: Toxicity; 9000: Microbiological examination; 10000: Biological examination.

7.040

United States. Environmental Protection Agency. Office of Pesticide Programs. Analytical Chemistry Branch. *Manual of chemical methods for pesticides and devices*. 2nd ed. [newly rev.] Arlington, VA: Published and distributed by AOAC International, 1992- . 1 v., looseleaf. 0935584471 SB951 668.65

Contains 284 methods for analysis of 190 chemicals commonly used in commercial pesticide formulations and technical materials. There are 18 new methods; outdated and redundant methods are eliminated, as well as those for which an equivalent procedure exists in *Official Methods of Analysis of the AOAC*. Entries include chemical names and synonyms, formulas, physical properties, reagents, equipment, and procedure to be used for determination of the chemical. Index contains cross-references and chemical, common, and trade names; there are also a pesticide formulations bibliography and a conversion table for first and second edition method numbers.

SPECIAL TOPICS IN ENVIRONMENTAL CHEMISTRY

Pesticides and Agrochemicals

7.041

Ashgate handbook of pesticides and agricultural chemicals, ed. by G.W.A. Milne. Aldershot, England; Burlington, VT: Ashgate, 2000. 206 p. 0566083884 SB9951 631.8 00-102728

This handbook describes 1,813 substances, including mixtures widely used in agriculture. Most pure chemicals include appropriate CASRNs and associated EINECS (European Inventory of Existing Chemical Compounds Subnumber). This is followed where possible by the American and European identification numbers and also the Merck Index number. Entries may contain information on chemical composition, function, application, physical properties, and suppliers; properties include melting point, boiling point, density or specific gravity, refractive index, optical rotation, ultraviolet absorption, solubility, and acute toxicity. The entries are divided into 12 subunits: acaricides, agricultural chemicals, animal feeds, fertilizers, fungicides, herbicides, insecticides, molluscicides, nematicides, plant growth regulators, rodenticides, and slimicides. The information is listed alphabetically in dictionary format. Synonyms or trade or trivial names are also listed and indexed; and companies supplying the products are given as well. Several indexes help locate specific products. Glossary; abbreviation section.

7.042

Briggs, Shirley A., and the staff of Rachel Carson Council. *Basic guide to pesticides: their characteristics and hazards*. Washington: Hemisphere Pub. Corp., 1992. 283 p. 1560322535 RA1270.P4 363.17 92-7024

Provides more than 700 listings for pesticides and their known transformation products. Tables summarize principal applications, persistence in the environment, acute and chronic mammalian toxicity, and toxicity to other organisms. Index cross-referenced by common name, chemical name, and CASRN.

7.043

CRC handbook of pesticides, ed. by G.W.A. Milne. Boca Raton, FL: CRC Press, 1995. 402 p. 0849324475 SB961 615.9 94-39758

Lists 386 most commonly used pesticides registered with EPA. Lists physical properties, acute toxicity, structural and molecular formulas, CASRN, and Merck Index number. Offers English and foreign names, trivial names, synonyms, and trade names.

7.044

Handbook of pesticide toxicology, ed. by Robert I. Krieger. 2nd ed. San Diego: Academic Press, 2001. 2 v., 1,908 p. 0124262600 (set) RA1270.P4 615.9 2001-89145

V. 1: Principles; v. 2: Agents. Previous ed. edited by Wayland J. Hayes Jr. and Edward R. Laws Jr.

7.045

Montgomery, John H. *Agrochemicals desk reference*. 2nd ed. Boca Raton, FL: CRC Press, 1997. 656 p. 1566701678 TD196.A34 628.5 97-77

This new edition adds additional chemicals and new information on their fate and transport in various environments. Some 200 chemicals are listed alphabetically according to EPA nomenclature, and, together with synonyms, are cross-referenced. Structural formulas, CASRNs, Department of Transportation designation, molecular formulas, formula weights, and Registry of Toxic Effects of Chemical Substances (RTECS) numbers are given. Also provided are physical and chemical properties, appearance and odor, bioconcentration factor, boiling point, dissociation constant, Henry's Law constant, hydrolysis half-life, ionization potential, soil/sediment partition coefficient, melting point, octanol/water partition coefficient, photolysis half-life, solubility in organics, solubility in water, specific density, environmental fate, vapor density, and vapor pressure. Fire hazards are listed, including flash point, and lower and upper explosion limits. Included are health hazard data, such as immediate danger to life or health (IDLH), and exposure limits such as time-weighted average (TWA), threshold limit value (TLV), short-term exposure limit (STEL), and TLV-ceiling; formulation types, toxicology information, and uses. Useful appendixes contain, among other items, EPA approved text methods, empirical formula index, bulk density values and porosity values for selected soil and rocks, solubility data, and synonym and cumulative indexes.

Air Pollution and Atmospheric Chemistry

7.046
Berner, Elizabeth Kay, and Robert A. Berner. *Global environment: water, air, and geochemical cycles.* Upper Saddle River, NJ: Prentice Hall, 1996. 376 p. 0133011690 QC880.4.A8 551.5 95-25174

Addresses areas of environmental concern occurring on a global or a regional scale, including water and energy cycles; atmospheric, rainwater, and water chemistry; and the processes and problems in rivers, lakes, and oceans. Quantitative estimates of effects are presented extensively throughout.

7.047
Blewett, Stephen E., with Mary Embree. *What's in the air: natural and man-made air pollution.* Ventura, CA: Seaview Pub., 1998. 140 p. 0964056526 TD883 363.739 98-70936

Moves from the environmental view to the industrial or economic view without espousing either, and discusses air pollution sources such as volcanoes, tornadoes, auto emissions, industrial emissions, etc., in significant detail. Treats smog history (natural and man-made), regulations, and mechanisms of formation and both sides of the ozone depletion problem. It ends with a chapter on the possible solutions to air pollution problems and the difficulties encountered with each of these proposed solutions. A remarkably complete short work on air pollution. Useful glossary.

7.048
Butler, John D. *Air pollution chemistry.* London; NY: Academic Press, 1979. 408 p. 0121479501 TD883 628.5 78-18677

Treats health factors, pollutant sources and removal from the atmosphere, sampling techniques and analytical methods, atmospheric chemistry, meteorological aspects of pollutant dispersion, and urban atmospheres.

7.049
Miller, E. Willard, and Ruby M. Miller. *Indoor pollution: a reference handbook.* Santa Barbara, CA: ABC-CLIO, 1998. 330 p. (Contemporary world issues) 0874368952 TD883 363.739 98-25606

A summary is followed by sections on characteristics, health effects, standards, sources, and control. The section on sources (52 pages) offers at least 35 different chemical and biological causes of human illness and discomfort in an indoor environment. Chapter bibliographies have hundreds of citations, including journal articles and audiovisuals; some individual pollutants have their own reference lists. Glossary; acronyms; measurements; testing information.

7.050
Seinfeld, John H. *Atmospheric chemistry and physics: from air pollution to climate change.* NY: Wiley, 1998. 1,326 p. (A Wiley-Interscience publication) 0471178152; 0471178160 pbk QC879.6 551.5 97-76938

Successor to Seinfeld's *Atmospheric Chemistry and Physics of Air Pollution*, 1986. Substantially revised to reflect advances in understanding of complex dynamics of the atmosphere. New chapters treat cloud physics, stratospheric chemistry, wet deposition, acid rain, chemical transport models, and other topics.

Soil and Water Chemistry

7.051
Ball, Philip. *Life's matrix: a biography of water.* NY: Farrar, Straus, and Giroux, 2000. 0374186286 GB661.2 553.7 99-59110

A thought-provoking journey about the origin and nature of water. Full of fascinating facts drawn from different directions, it discusses such topics as water's connection to the origin and maintenance of life, water's role in Earth's early stages of formation, and the unusual and unique properties of water due in large part to hydrogen bonding. It concludes with a very practical chapter on water supply and consumption, "Blue Gold." The bibliography lists excellent references and interesting anecdotes.

7.052

Evangelou, V.P. *Environmental soil and water chemistry: principles and applications.* NY: Wiley, 1998. 564 p. (A Wiley-Interscience publication) 0471165158 TD878 628.5 98-13433

An excellent, advanced book and resource in environmental chemistry, presenting physical and analytical chemistry and applications useful for both environmental engineers and chemists. Applications discuss both problems and solutions based on up-to-date models and state-of-the-art technologies. Fundamental topics include acid-base equilibria, hydrolysis of salts, complexes, electrochemistry, reaction kinetics, solubility, and mineral chemistry; applications include decomposition of organic wastes, synthetic organic chemicals, adsorption, colloidal systems, pesticides and heavy metals, and removal and neutralization procedures for select chemical contaminants. Well-presented charts and graphs; clear diagrams; efficient, easy-to-understand tables; clear, standard notation (SI units) throughout.

7.053

Franks, Felix. *Water: a matrix of life.* 2nd ed. Cambridge: Royal Society of Chemistry, 2000. 225 p. (RSC paperbacks) 085404583x QD169.W3 553.7

Update and extension of the 1982 edition. Discusses current scientific knowledge of water, properties, influence on dissolved substances, and role in life sciences and ecology. Treats water quality, usage, economics, and politics.

7.054

Montgomery, John H. *Groundwater chemicals desk reference.* 3rd ed. Boca Raton, FL: CRC Lewis Publishers, 2000. 1,345 p. 1566704987 TD426 628.1 00-37077

Contains numerous listings for hundreds of chemicals together with environmental fate listings. Emphasizes organic compounds found in groundwater, surface water, soil, air, and plants. Profiles all US EPA organic priority pollutants and compounds most commonly found in the workplace.

7.055

Pankow, James F. *Aquatic chemistry concepts.* Chelsea, MI: Lewis Publishers, 1991. 673 p. 0873711505 GB855 551.46 91-18535

Provides an overview of acid-base, mineral and solution, metal-ligand, and redox chemistry, as well as the effects of electrical charges on solution chemistry.

7.056

Stumm, Werner. *Chemistry of the solid-water interface: processes at the mineral-water and particle-water interface.* NY: Wiley, 1992. 428 p. (A Wiley-Interscience publication) 0471576727 GB855 551.46 92-9701

Reviews the scope of aquatic surface chemistry; coordination chemistry of hydrous oxide-water interface; surface charge and electric double layer; adsorption; kinetics of surface controlled dissolution of oxide minerals: weathering; precipitation and nucleation; particle-particle interaction; carbonates; redox processes; heterogeneous photochemistry; and trace elements and the solid-water interface in surface waters. Chap. 11 by Laura Sigg; Chap. 10 by Barbara Sulzberger.

7.057

Stumm, Werner, and James J. Morgan. *Aquatic chemistry: chemical equilibria and rates in natural waters.* 3rd ed. NY: Wiley, 1996. 1,022 p. (A Wiley-Interscience publication) 0471511846; 0471511854 pbk GB855 359.9 94-48319

Presents a complete introduction to concepts, applications, and techniques of aquatic chemistry. Treats chemical thermodynamics and kinetics; acid-base chemistry; dissolved carbon dioxide; atmosphere-water interactions; metal ions in aqueous solution and aspects of coordination chemistry; precipitation and dissolution; oxidation-reduction equilibria; solid-solution interfaces; trace metals: cycling, regulation, and biological roles; kinetics of redox processes; photochemical processes; kinetics of solid-water interface: adsorption, dissolution of minerals, nucleation, and crystal growth; particle-particle interaction: colloids, coagulation, and filtration; regulation of the chemical composition of natural waters; and thermodynamic data.

7.058

Van der Leeden, Frits, Fred L. Troise, and David Keith Todd. *The water encyclopedia.* 2nd ed. Chelsea, MI: Lewis Publishers, 1990. 808 p. (Geraghty & Miller ground-water series) 0873711203 TD351 553.7 89-14011

1st ed. by David Keith Todd. Includes facts and data on groundwater contamination, drinking water, floods, waterborne diseases, global warming, climate change, irrigation, water agencies and organizations, precipitation, oceans and seas, rivers, lakes, waterfalls, water use/reuse, snow and snowmelt, and hydrologic elements.

Metals

7.059

Katz, Sidney A., and Harry Salem. *The biological and environmental chemistry of chromium*. NY: VCH, 1994. 214 p. 1560816295 RA1231 574.19 94-5417

Chromium is an essential element and is responsible for normal glucose, carbohydrate, and fat metabolism. It is considered as a cofactor for insulin. However, chromium in the +6 oxidation state is a human carcinogen. Environmental problems as well as policy associated with the use of chromium are evaluated in chemical detail.

Hazardous Materials Handling

7.060

Blackman, William C. Jr. *Basic hazardous waste management*. 3rd ed. Boca Raton, FL: Lewis Publishers, 2001. 468 p. 1566705339 TD1040 363.72 2001-20391.

Fourteen chapters treat topics, impacts, technologies, problems, and issues relevant to conventional hazardous wastes, and management practices governed by statutes and regulations. Includes management of medical and infectious wastes, radiological wastes, and underground storage tanks.

7.061

Burgess, William A. *Recognition of health hazards in industry: a review of materials and processes*. 2nd ed. NY: Wiley, 1995. 538 p. (A Wiley-Interscience publication) 0471577162 RC967 363.11 94-18498

Identifies major health issues faced by workers in industry. Includes chapters on plastics fabrication, microelectronics, and chemical processing, as well as other topics covered in previous versions, such as aluminum, iron, and steel production; abrasive blasting; acid and alkali cleaning; degreasing; grinding, polishing, and buffing; forging; foundry operations; machining; welding; heat treating; nondestructive testing; electroplating; metal thermal spraying; painting; petroleum refinery operations; rubber production; acids, ammonia, and chlorine; paint and coatings; soldering of electrical products; batteries; quarrying; mining; smelting; asbestos products; asphalt products; abrasives; glass; ceramics; pulp and paper; and textiles.

7.062

Burke, Robert. *Hazardous materials chemistry for emergency responders*. Boca Raton, FL: CRC Press, 1997. 348 p. 1566701740 T55.3.H3 604.7 96-31459

Introduces chemistry to those frequently encountering hazardous chemicals but who have little or no formal chemistry background. The target audience includes practitioners and emergency response trainers providing short courses. The first chapter introduces physical and organic chemistry. Although not a substitute for basic undergraduate courses, the book highlights many basic principles. Each of nine chapters describes a specific class of chemical hazard, e.g., explosives, flammable liquids and solids, oxidizers, poisons, corrosives, etc. These chapters describe the characteristics of the hazard class, some of the top industrial chemicals, and both success stories and disasters in dealing with chemical accidents.

7.063

Chemical safety matters. International Union of Pure and Applied Chemistry [and] International Programme on Chemical Safety. Editorial group, Peter H. Bach [et al.]. Cambridge; NY: Cambridge University Press, 1992. 284 p. 0521413753 QD63.5 542 91-28312

Reviews safe use and proper disposal of hazardous chemicals in labs, and includes transportation, landfill protocols, incineration, recycling, explosives, and other sensitive subjects.

7.064

Lunn, George, and Eric B. Sansone. *Destruction of hazardous chemicals in the laboratory*. 2nd ed. NY: Wiley, 1994. 501 p. (A Wiley-Interscience publication) 047157399x TD1050 604.7 93-35634

Provides very detailed procedures for safe destruction of many dangerous chemicals using methods to reduce costs, hazards, and wastes.

7.065

Manahan, Stanley E. *Hazardous waste chemistry, toxicology, and treatment*. Chelsea, MI: Lewis Publishers, 1990. 378 p. 0873712099 TD1030 628.4 90-33889

Introduces chemistry and provides fundamental properties of hazardous materials (inorganic, organic, and biological); and surveys waste management strategies, providing recycling, treatment, and disposal options.

7.066

Rapid guide to hazardous chemicals in the workplace, ed. by Richard J. Lewis, Sr. 4th ed. NY:

Wiley-Interscience, 2000. 261 p. 0471355429 T55.3.H3 604.7 99-86808

A brief pocket guide to dangerous properties of most-frequently encountered industrial materials (some 750) cross-referenced to the 10th ed. of *Sax's Dangerous Properties of Industrial Materials*.

7.067

Wentz, Charles A. *Hazardous waste management.* 2nd ed. NY: McGraw-Hill, 1995. 580 p. (McGraw-Hill chemical engineering series) 0070693080 TD1030 363.72 95-3996

Comprehensive coverage of hazardous waste management field. Presents basic scientific and engineering principles together with pertinent legislation and case studies. Treats risk assessment, environmental legislation, and specifics of TSCA, RCRA, and Superfund Amendments and Reauthorization Act (SARA). Discusses hazardous waste characteristics, waste minimization practices, resource recovery, and treatment using physical, chemical, and biological methods; and mentions groundwater contamination, landfills, injection wells, facility selection and siting, and site remediation.

TOXICOLOGY

7.068

Comprehensive toxicology; eds. in chief, I. Glenn Sipes, Charlene A. McQueen, and A. Jay Gandolfi. NY: Pergamon, 1997-2002. 14 v. 0080423019 (set) RA1199 615.9 2001-55706

V. 1: General principles; v. 2: Toxicological testing and evaluation; v. 3: Biotransformation; v. 4: Toxicology of the hematopoietic system; v. 5: Toxicology of the immune system; v. 6: Cardiovascular toxicology; v. 7: Renal toxicology; v. 8: Toxicology of the respiratory system; v. 9: Hepatic and gastrointestinal toxicology; v. 10: Reproductive and endocrine toxicology; v. 11: Nervous system and behavioral toxicology; v. 12: Chemical carcinogens and anticarcinogens; v. 13: Index; v. 14: Cellular and molecular toxicology (publ. in 2002 by Elsevier).

7.069

Crosby, Donald G. *Environmental toxicology and chemistry.* NY: Oxford University Press, 1998. 336 p. (Topics in environmental chemistry) 0195117131 RA1226 615.9 97-22438

Summarizes environmental toxicology; some chapters assume a reasonably thorough background in biochemistry, others in organic chemistry; yet others expand concepts usually introduced in engineering. This, however, is the nature of environmental toxicology and chemistry. References to recent papers and books at the end of chapters are helpful; special topics at the end of each of the 16 chapters are welcome extensions, as they often cover items of recent interest, e.g., oil spills, chemical carcinogenesis, and chlorination.

7.070

Fenton, John Joseph. *Toxicology: a case-oriented approach.* Boca Raton, FL: CRC Press, 2002. 573 p. 0849303710 RA1219 2001-35682

Successfully discusses toxicology from the perspective of both clinical and analytical chemistry, resulting in a treatment of the principles of general toxicology for both clinicians and analytical scientists. Fenton thoroughly explains toxicology as applied to major organ systems; comprehensively covers chemistry, diagnoses, and treatment of specific toxins; and describes analytical strategies for lab testing of drugs and poisons. This book relies on case studies to stimulate interest; these studies were carefully selected for their abilities to teach important principles related to toxicology. Many interesting anecdotes related to the history of toxicology are included. A thorough and comprehensive work.

7.071

Manahan, Stanley E. *Toxicological chemistry and biochemistry.* 3rd ed. Boca Raton, FL: Lewis Publishers, 2003. 425 p. 1566706181 RA1219.3 815.9 2002-72486

Outlines basic concepts of general and organic chemistry and biochemistry as background to the topics in the book. Presents an overview of environmental chemistry, and discusses biodegradation, bioaccumulation, and biochemical processes in water and soil. Material on toxic effects considers toxicities to endocrine and reproductive systems, xenobiotic effects, and genetic toxicology—newer topics to this edition.

7.072

Ottoboni, M. Alice. *The dose makes the poison: a plain-language guide to toxicology.* 2nd ed. NY: John Wiley and Sons, Inc., 1997. 244 p. 0471288373 RA1213 91-7239

Reviews toxicology for scientists and managers throughout industry, as well as public health personnel and anyone with a need to know about toxic materials.

7.073

Patty's industrial hygiene; Robert L. Harris, ed.; contributors, H.E. Ayer ... [et al.]. 5th ed. NY: Wiley, 2000- . v. <1-3 > (A Wiley-Interscience publication) 0471297844 (set); 0471297569 (v. 1); 0471297542 (v. 2); 0471297534 (v. 3); 0471287496 (v. 4) RC967 613.6 99-32462

A new edition (4th ed., 1991-95) of the industrial hygiene portion of *Patty's Industrial Hygiene and Toxicology*. V. 1, I: Introduction to industrial hygiene; II: Recognition and evaluation of chemical agents; v. 2, III: Physical agents; IV: Biohazards; V: Engineering control and personal protection; v. 3, VI: Law, regulation, and management; v. 4, VII: Speciality areas and allied professions.

7.074

Patty's industrial hygiene and toxicology. 4th ed. (some volumes are 3rd ed.) Ed. by George D. Clayton and Florence E. Clayton; contributors, R.E. Allan [et al.]. NY: Wiley, 1991-<1995> 3 vol. in 10 <v.1, pts. A-B; v.2, pts. A-F; v.3, pts. A-B > (A Wiley-Interscience publication) 0471501972 (v.1, pt.A); 0471501964 (pt.B) RC967 613.6 90-13080

V. 1: General principles, Parts A-B: Discusses all aspects of industrial hygiene, including air pollution, legislation, reports and records, fire and explosion hazards, sampling, analysis, and calibration. V. 2: Toxicology: Part A Includes a historical overview of toxicology, and has chapters on phenols, esters, aldehydes, and other compounds; Part B Reviews occupational carcinogenesis, aliphatic and aromatic hydrocarbons, and more; Part C Updates the toxicology of glycols, nitrites, cyanides, and others. Part C includes aerospace medicine, and an updated section on ketones and metals. Part D treats carbon black, occupational medical surveillance, halogenated cyclic hydrocarbons, alcohols, esters, cyanides, nitriles, glycol ethers, dioxins, PCBs, and bridged diphenyls. Part E includes heterocyclic and miscellaneous nitrogen compounds, aliphatic carboxylic acids, synthetic polymers, and halogenated aliphatic hydrocarbons. Part F reviews diagnoses of occupational and environmental diseases, flame retardants, and inorganic compounds of carbon, nitrogen, and oxygen. In addition, there is included material on organic sulfur compounds, boron, the halogens, and glycol. V. 3: Theory and rationale of industrial hygiene practice. Part A, The work environment. Covers health surveillance programs, occupational health nursing, and detection of occupational diseases; Part B, Biological responses. Includes ergonomics, radiation, and noise exposure.

7.075

Sittig's handbook of toxic and hazardous chemicals and carcinogens. 4th ed. Richard P. Pohanish, ed. Norwich, NY: Noyes Publications, 2002. 2 v., 2,608 p. 081551459x RA1193 615.9 2001-56289

Rev. of *Handbook of Toxic and Hazardous Chemicals and Carcinogens*, by Marshall Sittig, 3rd ed., 1991. A compendium on nearly 1,500 hazardous chemicals, the results of years of compilation. After brief introductory material, chemicals are listed alphabetically across the two volumes. Summarizes currently available information under chemical description and synonyms, potential for exposure and permissible exposure limits, toxicity, medical surveillance, first aid, protective methods, spill and fire cleanup, and disposal methods. Other information includes regulatory authorities that have set standards, and states with relevant regulations. Sections on permissible and recommended exposure levels and determination in air and water. Entry details vary depending on information available for particular chemicals. Information appears to be accurate and comprehensive. Eight appendixes discuss oxidizers, carcinogens, and various cross-indexing information. These volumes are a valuable resource, although much of the material is similar to that in the third edition (1991).

7.076

Sterner, Olov. *Chemistry, health, and environment.* Weinheim; NY: WIley-VCH, 1999. 345 p. 3527300872 RA1226 615.902

Discussion is primarily about the effects of chemicals on humans; there is relatively little about environmental effects. About a third of the book (six of 13 chapters) discusses relatively basic concepts: organic chemistry (e.g., structures, properties); biochemistry (e.g., cells, nucleic acids); toxicology (e.g., dose-response relations); anatomy (e.g., functions of organs). Some of the other chapters are quite sophisticated, such as "Metabolism of Exogenous Compounds in Mammals" and "The Molecular Basis for Genotoxicity and Carcinogenicity." Comprehensive glossary. Contains explanations for toxic effects of many chemicals currently of interest: thalidomide, methanol, aflatoxin, nitrates, and poison ivy toxins.

7.077

Thornton, Joe. *Pandora's poison: chlorine, health, and a new environmental strategy.* Cambridge, MA: MIT Press, 2000. 599 p. 0262201240 RA1242.C436 615.9 99-57011

Thornton, a biologist, urges that use of all organochlorine

chemicals, not just specific organochlorines deemed problematic by risk assessment procedure, be phased out. Organochlorine pollution worldwide, organochlorine toxicity, and danger to wildlife and people are discussed. Considers production of the parent compound, chlorine, its necessary tie-in to alkali production, and the production of organochlorines, which produces many byproducts including dioxins.

Regulatory and Legislative Concerns

7.078
Educating for OSHA savvy chemists, ed. by Paul J. Utterback and David A. Nelson. Washington, DC: American Chemical Society, 1998. 191 p. (ACS symposium series, ISSN 0097-6156; 700) 0841235694 QD63.5 542 98-16896

Addresses the dual issues of curriculum reform as well as compliance with laboratory health and safety regulations, and recommends creating opportunities to integrate chemical health and safety education into existing courses. Up-to-date, precise explanations of chemical health and safety terms are provided; relevant OSHA regulations are clearly reviewed, explained, and summarized; practical strategies are presented and justified to make it easy to comply with these codes and regulations. Up-to-date references and resources, including Internet addresses.

7.079
NIOSH pocket guide to chemical hazards. Cincinnati, OH: U.S. Dept. of Health and Human Services, Public Health Service, Centers for Disease Control and Prevention, NIOSH; Washington, DC: for sale by the Supt. of Docs., US G.P.O, 1997. 440 p. (DHHS [NIOSH] publication no. 97-140) T55.3

Small, "pocket-sized" spiral-bound book providing, in tabular form, chemical and physical properties, PELs, Immediately Dangerous to Life or Health (IDLH) levels, health hazards, chemical incompatibilities, personal protection recommendations, and first aid procedures.

7.080
NIOSH pocket guide to chemical hazards and other databases [electronic resource]. [Cincinnati]: U.S. Dept. of Health and Human Services, Public Health Service, Centers for Disease Control and Prevention, National Institute for Occupational Safety and Health, [2000]. 1 CD-ROM (DHHS [NIOSH] publication; no. 2000-130)

Contains 8 different databases on chemical health and safety: Immediately Dangerous to Life and Health (IDLH) Concentrations, International Chemical Safety Cards (ICSDC), NIOSH Manual of Analytical Methods, NIOSH Pocket Guide to Chemical Hazards, NIOSH Recommendations for Protective Clothing, Specific Medical Tests for OSHA Regulated Substances, Toxicological Reviews of Selected Chemicals, and the 1996 North American Emergency Response Guidebook (US DOT). IDLH lists 387 substances in alphabetical order; ICSDC cross-references chemical substances by CASRN, RTECS number, and UN number. The Pocket Guide lists synonyms; flammability, toxicity, and reactivity ratings; physical data; and PEL values; Analytical Methods lists substances in alphabetical order by four-digit code; Protective Clothing, for handling of or exposure to selected hazardous substances, lists substances alphabetically, identifies primary route of entry, and offers protection suggestions. OSHA- regulated chemicals includes specific medical tests conducted and their results, its authors or editors, and references with dates; Toxicological Review of Selected Chemicals contains name and CASRN; tests provided excellent articles and research studies. OSHA PEL/TWA, STEL, and CEILING values are included. DOT emergency guide is cross-referenced and can be searched by name, UN number, commercial vendor, etc., with points of caution color-coded.

7.081
Registry of toxic effects of chemical substances (RTECS). Cincinnati, OH: U.S. Department of Health and Human Services, Public Health Service, Center for Disease Control, NIOSH; Washington, DC: for sale by the Supt. of Docs., US G.P.O., 1975-1986. RA1215 94-27926

Available as an online service (see under Electronic Databases) and in microfiche, 1977-1993. This carcinogen data source provides for more than 90,000 chemicals their CASRNs, molecular formulas, reproductive effects, federal standards, EPA status, IARC reviews, and such effects as irritation.

Environmental Chemistry Database

7.082
TOXLINE on SilverPlatter. Boston: SilverPlatter Information, Inc., 1989- . CD-ROM. Quarterly; cumulative discs; online access via Dialog. Z7890x

Contains the TOXLINE file, part of the National Library of Medicine's online bibliographic family of databases, as well as certain other related documents and databases.

Chapter 8

Industrial Chemistry

In this chapter, the applied field of industrial chemistry will be reviewed. Here will be found general resources in the chemical industrial field, which touches on chemical engineering; later portions of the chapter discuss special and applied topics that fall within the industrial subject area. The items listed are by no means exhaustive; they are only a fraction of the rich resources available to the industrial chemist and chemical engineer. In addition, some of the resources that would normally belong in this installment have already been included in previous chapters.

If the reader remembers, in the very first chapter, the organization of library collections according to the Dewey Decimal and Library of Congress Classification systems was mentioned. Chemical engineering, 660 in the Dewey system, is organized as follows:

- 660 — chemical engineering
- 661 — industrial chemicals
- 662 — explosives
- 663 — beverages
- 664 — food technology
- 665 — oils, fats, waxes, and gases
- 666 — ceramic and allied industries
- 667 — cleaning and dyeing
- 668 — other organic products
- 669 — metallurgy

The Library of Congress (LC) System has a similar breakdown for T, technology, organized thus:

- T   — technology - general
- TA — engineering - general
- TC — hydraulic engineering
- TD — sanitary and municipal engineering
- TE — roads and pavements
- TF — railroad engineering and operation
- TG — bridges and roofs
- TH — building construction
- TJ — mechanical engineering and machinery
- TK — electrical engineering and industries
- TL — motor vehicles; cycles; aeronautics
- TN — mineral industries; mining and metallurgy
- TP — chemical technology
- TR — photography
- TS — manufactures
- TT — trades
- TX — domestic science (including cookery)

This chapter will be devoted to information sources from only a few parts of the 660 and the T classifications. From 660, we will concentrate on sources in 661, industrial chemicals; 665, oils, fats, waxes; 666, ceramic and allied industries; and 669, metallurgy. Borrowing from the T areas, we will mention T54-55.3, for laboratory safety; TA400s, where materials and composites are classed; TD 169-500, environmental pollution and protection, including water pollution; TD878-899, industrial wastes as well as air and soil pollution; TN, mineral industries, especially coal; TP, chemical technology; and TS800-1932, manufactures, where so many items are found that are termed "practical" or useful to the consumer. These cover leather, wood, adhesives, paper, petroleum, rubber, and plastics. These divisions will be used very loosely to organize the material in this installment. Within each of these areas will be found the usual divisions of encyclopedias, journals, dictionaries, handbooks, directories, and treatises, where suitable.

HANDBOOKS AND ENCYCLOPEDIAS

8.001

CRC materials science and engineering handbook
[ed. by] James M. Shackelford and William Alexander. 3rd ed. Boca Raton, FL: CRC Press, 2001. 1,949 p. 0849326966 TA403.4 620.1 00-48567

Structure of materials; Composition of materials; Phase diagram sources; Thermodynamic and kinetic data; Thermal properties of materials; Mechanical properties of materials; Electrical properties of materials; Optical properties of materials; Chemical properties of materials; Selecting structural properties; Selecting thermodynamic and kinetic properties; Selecting thermal properties; Selecting mechanical properties; Selecting electrical properties; Selecting optical properties; Selecting chemical properties.

8.002

LabGuide. Washington, DC: American Chemical Society, 1973- . Annual. ISSN 1520-4782 Q180.56 507 88-24133

This annual guide, published as the second August issue of *Analytical Chemistry* (1972/1973-), is a supplement to selected American Chemical Society publications (1997/1998-), sent to all subscribers of ACS journals; it contains much useful industrial information. Included are advertised products, product index, instruments and laboratory supplies, laboratory reagents and standards, services, original equipment manufacturers' supplies and suppliers, company and advertiser directories, and reader service information. Also available on the Internet (current edition).

8.003

Lide, David R. *Basic laboratory and industrial chemicals: a CRC quick reference handbook.* Boca Raton, FL: CRC Press, 1993. 370 p. 0849344980 QD64 546 93-25269

A convenient and inexpensive book with properties of most-used chemicals compounds, this work includes data for only about a thousand organic and inorganic chemicals; biochemical materials are excluded. Data include molecular weight, melting and boiling points, critical temperature and pressure, electric dipole moment, first ionization potential, phase transition properties, thermodynamic and transport properties, and unusual hazards.

8.004

Perry's Chemical engineers' handbook, 7th ed. Prepared by a staff of specialists under the editorial direction of late editor Robert H. Perry; editor, Don W. Green; assistant editor, James O. Maloney. New York: McGraw-Hill, 1997. various pagings. 0070498415 TP151 660 96-51648

A classic chemical engineers' handbook in 30 sections, with math tables; physical and chemical property data; thermodynamic data; fluid mechanics, transport, and storage of fluids data; transport, storage, and grinding of solids; heat transport and generation; heating, ventilation, and air conditioning; distillation, gas absorption, liquid extraction,

adsorption and ion exchange, and other separation processes; gas-liquid systems, liquid-solid systems, liquid-liquid, and solid-solid systems; and other information.

8.005

Perry's chemical engineers' andbook [computer file]. NY: McGraw-Hill, 1999. 7th ed. 1 CD-ROM TP151.

Electronic version of *Perry's Handbook*; see previous entry for details.

8.006

Riegel, Emil Raymond. *Riegel's handbook of industrial chemistry,* ed. by James A. Kent. 9th ed. NY: Van Nostrand Reinhold, 1992. 1,288 p. 0442001754 TP145 660 92-22660

Chapters by specialists on chemical process industries— economics, plastics, rubber, synthetics, fibers, dyes, synthetic organic chemicals, processing, raw materials, equipment, production figures, industrial wastewater treatment, air pollution abatement procedures.

8.007

Shreve, Randolph Norris. *Chemical process industries* (Shreve's Chemical process industries), ed. by George T. Austin. 5th ed. NY: McGraw-Hill, 1984. 859 p. 0070571473 TP145 660.2 83-13629

Discusses new legislation on environmental product safety in conjunction with chemical process industry. Contains production and sales statistics for all chemicals and process chemical reactions discussed. Contains new chapters on water conditioning, energy, fuels, food processing, and others of immediate importance.

8.008

Ullmann, Fritz. *Ullmann's encyclopedia of industrial chemistry.* 6th, completely rev. ed. Executive ed., Wolfgang Gerhartz; senior ed., Y. Stephen Yamamoto; editors, F. Thomas Campbell, Rudolf Pfefferkorn, James F. Rounsaville. Weinheim, Germany; Wiley-VCH, 2003. 40 v. 3527303855 TP9 660

The complete set includes reaction and process technology; process engineering; process development and design of chemical plants; physical analysis methods; measurement technology; environmental control and occupational safety; industrial compounds (inorganic, metals, organic); building materials, glass, ceramics;

natural gas, coal, petroleum; fibers, textiles, wood, paper, leather; drugs and cosmetics; pest control; foods; lasers; voltaic cells; emulsions; nuclear energy. Trademark list is included in every third volume of the last 18 volumes. CD-ROM version (1997). Editors listed are those from the 5th edition.

8.009

Ullmann, Fritz. *Ullmann's encyclopedia of industrial chemistry [electronic resource].* 6th rev. and enl. ed. Weinheim: Wiley-VCH, 2002. TP9

Revised ed. of the previous electronic version of Ullmann's; accessed via the Internet.

8.010

Yaws, Carl L. *Matheson gas data book.* 7th ed. Parsippany, NJ: Matheson Tri-Gas; NY: McGraw-Hill, 2001 982 p. TP761.C65 665.7 2001-30867

Useful data on industrial gases, many of which Matheson sells. Profiles 80 gases; full properties for 157 gases. Contains physical, thermodynamic, environmental, transport, safety, and health and related properties of gases of major importance. Includes references to the literature.

DICTIONARIES AND THESAURI

8.011

Ash, Michael and Irene Ash, compilers. *Chemical tradename dictionary.* NY: VCH Publishers, 1993. 529 p. 1560816252; 3527896252 (Germany) TP9 660 92-35154

A sourcebook for locating and identifying chemical trade name product lines in the international marketplace. Alphabetical list of some 14,000 chemical trade names were compiled through direct contacts with some 2,300 manufacturers. Coverage includes materials used in manufacturing—e.g., cosmetic and food additives, detergents, emulsifiers, resins, paint and textile additives. Each entry gives manufacturer, common chemical name, and a brief statement of function or application. The second section, a manufacturer's directory, lists company address with telephone, telex, and fax numbers.

8.012

Ashford's dictionary of industrial chemicals,

compiled by Robert D. Ashford. 2nd ed. London: Wavelength, 2001. 1,269 p. 095226742x TP200 661.003

Some information from 1st ed., 1994. Treats 6,800 products, including major commodity chemicals, plastics, intermediates, additives, and specialty chemicals. Alphabetic arrangement includes properties, production, and uses of each chemical.

8.013

Fachwörterbuch, Chemie und chemische Technik: Deutsch-Englisch, mit etwa 62 000 Wortstellen= Technical dictionary of chemistry and chemical technology; German-English: with about 62,000 entries. 4th thoroughly revised and enlarged ed. by Technische Universitat Dresden, Zentrum für Angewandte Sprachwissenschaft. Berlin: A. Hatier, 1992. 760 p. 3861170353 QD5 540.3

Based on its 1989 English-German counterpart, this authoritative reference adds nearly 2,000 terms for a total of 62,000 entries. Vocabulary is revised and enlarged in several areas—chemical technology, biotechnology, kinetics, coal chemistry, food chemistry, hydrochemistry, water conditioning, waste water treatment and more. Entries include variant spellings and indicate German masculine and feminine nouns. Cross-references are used for obsolete terms.

8.014

Industrial chemical thesaurus. 2nd ed. Compiled by Michael and Irene Ash. NY: VCH, 1992. 2 v.: 831 p., 456 p. 1560816155 Z695.1 025.4 92-22095

V. 1: Chemical to trade name reference; v. 2: Trade name to chemical cross reference. This thesaurus will locate a single chemical marketed under a variety of trade names. Included are nearly 40,000 international trade names of 6,000 generic chemicals. V. 1 lists generic chemicals alphabetically, with synonyms, CASRN, definition, classification, formula, properties, precautions, toxicity, and applications; information following lists trade name equivalents and products containing the chemical in question. V. 2 lists trade names alphabetically, with manufacturer and generic chemical name.

8.015

Noether, Dorit and Herman Noether. *Encyclopedic dictionary of chemical technology.* NY: VCH, 1993. 297 p. 0895733293; 3527266968 TP9 660 93-105705

Provides essential terminology of chemical technology, with

encyclopedic entries. Chemical engineering systems, processes, and operations are discussed in brief statements as well as detailed accounts; there are helpful tables and diagrams.

HISTORY AND DEVELOPMENT OF CHEMICAL INDUSTRY

8.016

Brandt, E.N. *Growth company: Dow Chemical's first century.* East Lansing: Michigan State University Press, 1997. 649 p. 0870134264 HD9651.9.D6 338.7 97-749

Built on carefully annotated company records and resources, and an especially rich collection of more than 100 oral histories, Dow's longtime company historian and archivist has written a view of what Herbert and Willard Dow created. Reinventing the bromine process and establishing newer, better methods for workhorse industrial chemicals, Dow (father and son) brought the fledgling US chemical industry onto the world stage. References, notes, annotations, and a list of 30 patents, from bromine and phenol to polymerization catalysts, starburst dendrimers, and highly absorbing polymers.

8.017

Cain, Gordon. *Everybody wins!: a life in free enterprise.* Philadelphia, PA: Chemical Heritage Press, 1997. 342 p. (The Chemical Heritage Foundation series in innovation and entrepreneurship) 0941901149 HD9651.95.C34 338.7 96-48448

The chemical industry drives the US economy, contributes favorably to the balance of payments, and provides a base of career jobs in research and development, manufacturing, and business and finance. Cain's lifetime association with chemistry and industry shows the special place of the chemical industry in the US, and his autobiography is a validation of that testament. Written with brevity, clarity, and organization, Cain offers personal and professional history, lively and informative. He lived (and still lives) in interesting times that included the Great Depression, WW II, and the emergence of the chemical industry in the decades following the war.

8.018

Hermes, Matthew E. *Enough for one lifetime: Wallace Carothers, inventor of nylon.* [Washington, DC]: American Chemical Society and the Chemical Heritage Foundation, 1996. 345 p. (History of modern chemical sciences, ISSN 1069-2452) 0841233314 QD22.C35 660 96-5560

Carothers was hired by DuPont in 1928 to lead a group doing basic research in organic chemistry. Under his guidance, within a few years, the synthetic rubber neoprene and the first fully synthetic fiber, nylon, were discovered; but before nylon was in production, Carothers had committed suicide. This biography is clearly written and comprehensible even to readers with little chemistry. It includes much new information derived from letters to and from Carothers and interviews with relatives and friends. Illustrated; complete bibliography and index.

8.019

Multhauf, Robert P. *The history of chemical technology: an annotated bibliography.* NY: Garland Pub., 1984. 299 p. (Bibliographies of the history of science and technology; v. 5) (Garland reference library of the humanities, vol. 348) 0824092554 Z7914.C4 016.66 82-48272

Reviews the technological development of applied chemistry from ancient times to the present, limited somewhat to Western Europe and North America. Some 1,530 items are cited, including significant primary works, monographs, periodical articles, and some obscure items as well. Languages are mostly English, German, and French, and date to 1981.

8.020

Smiley, Robert A., and Harold L. Jackson. *Chemistry and the chemical industry: a practical guide for non-chemists.* Boca Raton, FL: CRC Press, 2002. 165 p. 1587160544 pbk TP145 660 2001-52817

Discusses the jargon, concepts, and key concerns of chemistry and the chemical industry; presents concise information accessibly and in a user-friendly format; uses easy-to-understand methods of presentation, explains concepts without using mathematics and with little physical science; and provides insights to those entering the chemical profession.

PRODUCT SOURCES

8.021

Ash, Michael, and Irene Ash, compilers. *Chemical manufacturers directory of trade name products.* 2nd ed. Endicott, NY: Synapse Information Resources, 2000. 1,130 p. 1890595233 TP12 660 99-96850

Lists nearly all international chemical manufacturers and their trade name products. The print version is divided into two parts, an alphabetical listing of international chemical manufacturers with their products, and a listing of trade name products cross-referenced to manufacturer. The trade name section includes a brief descriptor for each product. More than 57,000 products are listed, updated annually.

8.022

Ash, Michael, and Irene Ash, compilers. *Specialty chemicals source book.* 2nd ed. NY: Synapse Information Resources, 2001. 2 v., 2,136 p. 1890595322 TP200 00-192312

Specialty chemicals are materials of economic value serving general or niche markets, and often are raw materials used in the manufacture of other end products. This compilation provides information about more than 8,000 specialty chemicals important in agriculture, electronics, food processing, the pharmaceutical industries, and many other fields. Part 1 has entries for chemicals alphabetically by chemical name; each entry includes identifying information such as CASRNs, generic chemical and trade names, and molecular formulae. Synonyms are cross-referenced to main entries. Properties, toxicology, precautions for storage and use, regulatory information, manufacturers, and distributors are listed. Part 2, a keyword function/application index, allows searching for chemicals important to particular markets. Part 3 lists manufacturers, about 4,000 companies worldwide, with addresses, telephone, fax, and e-mail. Part 4 consists of other indexes with cross-references to the main entries from numerical identifiers and chemical formulae. A glossary of chemical industry terminology; list of abbreviations.

ECONOMIC AND FINANCIAL INFORMATION

Monographic Sources

8.023

Dun and Bradstreet/Gale industry reference handbooks; Stacy A. McConnell, ed.; Linda D. Hall, assoc. ed. Detroit: Gale, 1998-1999. 7 v. 0787630403 (set); 0787630020 (v. 1); 0787630047 (v. 2); 0787630055 (v. 3); 0787638390 (v. 4); 0787630039 (v. 5) HG4907 338.7 00-500443

Compiles information collected from corporate and government sources about US chemical companies,

associations, consultants, trade contacts, and trade shows. Introductory articles overview of history, participants, and trends. Federal statistics and projections are presented in tables and graphs. Most of the handbook is listings of 4,159 chemical companies. Each entry provides an identifying Dun & Bradstreet number for the company, name, parent (where applicable), address, telephone, sales, employees, primary Standard Industrial Classification (SIC), brief description of its business activity, and name and title of highest officer. Other sections have statistics, rankings, and information on recent acquisitions and mergers. Listings of trade associations, industry consultants, trade information sources, and trade shows. Master index of company and organization names, individual names, SIC industry names, and terms; separate index of conversion tables from SIC codes to the emerging North American Industry Classification System (NAICS).

8.024

Kline guide to the U.S. chemical industry, ed. by William T. Eveleth. 5th ed. Fairfield, NJ: Kline, 1990. 1 v., various pagings HD9651.5 338.4 90-179302

(Revised edition of *The Kline Guide to the Chemical Industry*, 4th ed., 1980.) Covers Marketing and economics of basic and intermediate chemicals; polymers and additives; agricultural and fine chemicals; coloring agents; miscellaneous industrial chemicals; industry organizations and information sources; and contains a list of 465 leading US chemical companies. Includes bibliographic references.

Selected Periodical Source

8.025

Chemical & engineering news. Easton, PA: American Chemical Society, 1942- . Weekly (except last week of December). ISSN 0009-2347 TP1 660.5 41-2413

Also called *C&EN*. Weekly news magazine sent to all ACS members. Each issue contains articles on industry and government activities as well as information on the profession in general. Includes "Facts and Figures for the Chemical Industry," a twice-yearly (Spring and September) feature. Available online via the Internet to subscribers.

STANDARDS AND SPECIFICATIONS

8.026

American Society for Testing and Materials. *Annual book of ASTM standards*. Philadelphia: ASTM, 1970- . Annual compilation; 71 v. in 16 sections. ISSN 0192-2998 TA401 620.1 83-641658

Industry and laboratory standards for nearly all classes of materials. Includes analytical methods, mechanical methods, etc. for testing. Recognized worldwide. Successor to *Book of A.S.T.M. Standards, including tentative standards*. Philadelphia: The Society, 1939-[71].

TREATISES AND COMPREHENSIVE WORKS

8.027

Anastas, Paul T., and John C. Warner. *Green chemistry: theory and practice*. Oxford, UK: NY: Oxford University Press, 1998. 135 p. 0198502346 TP155 660 98-36292

Defines green chemistry; introduces green chemistry tools, principles, effects, feedstocks and starting materials; evaluates reaction types and methods of designing safer chemicals; shows examples of green chemistry; and discusses future trends.

8.028

Clausen, Chris A. III, and Guy Mattson. *Principles of industrial chemistry*. NY: Wiley, 1978. 412 p. (A Wiley-Interscience publication) 047102774x TP145 660.2 78-9450

Fourteen chapters treat the chemical industry; basic considerations; material accounting; energy accounting; chemical transport; heat transfer; kinetics; separation processes; instrumentation; process development; patents; economics; research; and process development case study.

8.029

Coulson, J.M., and J.F. Richardson. *Coulson & Richardson's chemical engineering,* with J.R. Backhurst and J.H. Harker. Oxford; Boston: Butterworth-Heinemann, 1996-<1997 >. v. <1, 5-6 >. 0750625570 pbk (v. 1); 0750626127 pbk (v. 5); 0750625589 pbk (v. 6) TP155 660 95-23187

V. 1: Fluid flow, heat transfer, and mass transfer (5th ed., 1996); v. 5: Solutions to the problems in Chemical engineering, volumes 2 and 3, by J.R. Backhurst and J.H. Harker (2nd ed., 1997); v. 6: Chemical engineering design, ed. by R.K. Sinnott (2nd ed., 1996).

8.030

Lancaster, Mike. *Green chemistry: an introductory text*. Cambridge: Royal Society of Chemistry, 2002. 310 p. (RSC paperbacks) 0854046208 TP155.2.E58 660.20286

Discusses principles and concepts of green chemistry; waste production, problems, and prevention; measurements and control of environmental performance; catalysis and organic solvents; renewable resources; alternative energy sources; and case studies.

8.031

Materials science and technology: a comprehensive treatment, ed. by R.W. Cahn, P. Haasen, and E.J. Kramer. Weinheim; NY: VCH, 1991-<1999 >. <v. 1, 2A, pt 1; 2B, pt 2; 3A, 5-7, 9, 10A, pt. 1; 11-15, 18-19 > 18 v. 3527268138 (set); 1560811900 (set); 352729449x; 3527268316 TA403 620.1 90-21936

V. 1: Structure of solids; material includes crystal structure, lattices, intermetallic compounds and phases, and crystal interfaces. V. 2A: Characterization of materials (pt. 1); v. 2B: Characterization of materials <pt.2>. V. 3A: Phase transformations in materials. V. 6: Plastic deformation and fracture of materials; material includes flow and work hardening, deformation, stress-strain, creep, high-temperature deformation, plasticity, fracture, fatigue, fracture mechanics, friction, and wear. V. 7: Constitution and properties of steel. V. 9: Glasses and amorphous materials. V. 10A: Nuclear materials <pt. 1>. V. 11: Structure and properties of ceramics. V. 12: Structure and properties of polymers. V. 13: Structure and properties of composites, ed. by Tsu-Wei Chou. V. 14: Medical and dental materials. V. 15: Processing of metal and alloys. V. 18: Processing of polymers. [V. 19]: Synthesis of polymers.

8.032

Wittcoff, Harold A., and Bryan G. Reuben. *Industrial organic chemicals*. NY; Chichester: Wiley, 1996. 531 p. 0471540366 TP247 661.8 95-35580

Updated, expanded, and rewritten version of v. 1 of a two-volume set, *Industrial Organic Chemicals in Perspective*, 1980. This new version describes a mature chemical industry, and discusses problems such as over- and underproduction of certain chemicals, and difficulties generated by government regulations. International aspects of the chemical industry are also treated. The bulk of the book includes the industrial chemistry of the seven most important organic chemicals: methane, ethylene, propylene, the C4 hydrocarbons, benzene, toluene, and the xylenes. Polymers and industrial catalysts are also discussed.

Special Topics in Industrial Chemistry

Chemical Technology

Handbooks and Data Compilations

8.033

Archer, Wesley L. *Industrial solvents handbook.* NY: M. Dekker, 1996. 316 p. 0824797183 TP247.5 661.8 96-1568

Discusses solvent safe handling, and includes physical and chemical properties, uses, and toxicity as applied to chemical industry. Solvents are grouped by chemical functions. A companion computer disk includes spreadsheets.

8.034

Ash, Michael, and Irene Ash, compilers. *Handbook of industrial surfactants: an international guide to more than 21,000 products by trade name, composition, application, and manufacturer.* 2nd ed. Aldershot, Hampshire, England; Brookfield, VT: Gower, 1997. 2 v., 2,403 p. 0566078929 (set) TP994 668 96-29925

More than 16,000 trade name surface-active agents are listed. Contains trade name reference, trade name applications cross-reference, chemical component cross-reference, and manufacturers' directory.

8.035

Handbook of physical properties of organic chemicals, ed. by Philip H. Howard and William M. Meylan; associate ed., Julie Funk. Boca Raton, FL: Lewis Publishers, 1997. 1 v., various pagings. 1566702275 QD257.7 547 96-51427

Physical and chemical property data are included for about 13,000 compounds of environmental, pharmaceutical, and commercial interest. Entries are arranged in order of CASRN and include structure, formula, molecular weight, melting and boiling points, water solubility, octanol-water partition coefficient, vapor pressure, acid dissociation constant, Henry's Law constant, and atmospheric hydroxyl radical reaction rate constant. Some values are estimated or extrapolated. Chemical formula and name indexes.

8.036

Industrial solvents handbook, ed. by Ernest W. Flick. 5th ed. Westwood, NJ: Noyes Data Corp., 1998. 963 p. 0815514131 TP247.5 661 98-5137

A revised and expanded work (3rd ed., 1985; 4th ed., 1991), mainly designed for the chemical processing industry but useful for anyone needing data on industrial solvents. Helps selection of new solvents when original solvents are no longer available, environmentally inappropriate, or economically impractical. Some 1,200 tables with basic data on physical properties and solubilities; phase diagrams for multicomponent systems; and HPLC and UV data for a variety of solvents. Material in the tables was selected from manufacturers' literature and are referenced following the title. Complete references; trade name index.

Treatises

8.037

Chemistry of advanced materials: an overview, ed. by Leonard V. Interrante and Mark J. Hampden-Smith. NY: Wiley-VCH, 1998. 580 p. 0471185906 TA403 620.1 97-11898

Covers all major materials areas, such as electronic, magnetic, electroactive, optical, or biomaterials, and discusses metal oxides and nonoxides as well as polymers. Describes different synthesis techniques, e.g., precipitation in solution, gas phase synthesis, or polymeric precursor decomposition. Nanostructured or layered materials are included in various aspects in several chapters. Topics are discussed in great detail, yet still there is sufficient introduction provided. Extensive chapter references, up to 1995. Graphs and tables.

8.038

McDonough, William, and Michael Braungart. *Cradle to cradle: remaking the way we make things.* NY: North Point Press, 2002. 193 p. 0865475873 pbk TD794.5 745.2 2001-44245

The play on words in the title is the key to understanding the environmentalist authors' point of view and inspiring approach toward intelligent design for products and processes. Their pitch is to be good, not less bad, in our designs. American architect McDonough and German chemist Braungart are partners in a company they founded that advises corporations and institutions on eco-effective practices requiring no postmanufacturing and postconsumer problem-solving because there will be no problems.

8.039

Murray, G.T. *Introduction to engineering materials: behavior, properties, and selection.* NY: Dekker, 1993. 669 p. (Materials engineering; 2) 0824789652 TA403 620.1 93-12660

This is a work on the comparative classification of

materials and their processing, selection, and applications. Chapters treat classification of materials; properties, structure, crystal imperfections, diffusion, and plastic deformation; processing and properties of metals, ceramics, polymers, and composites; electronic materials; and environmental degradation of materials.

Minerals and Metals

8.040
Kurlansky, Mark. *Salt: a world history*. NY: Walker & Company, 2002. 484 p. 0802713734 TN900 553.6 2002-391600

The history of the world is told piquantly in this well-seasoned account of events that shaped world history and that often revolved around salt. Salt is deeply intertwined with the history of humankind because of salt's roles as a basic physiological requirement, as a preservative of food, and as a player in the holy-grail quest for flavor-packed food. From pickles to salt cod, from the wealth of Venice to fish sauce, from the Via Salaria to the Hanseatic League, from above-ground salt lakes and salt marshes to salt ponds, from Salzburg to Morton's, this book recounts it all.

8.041
Laszlo, Pierre. *Salt: grain of life*, transl. by Mary Beth Mader. NY: Columbia University Press, 2001. 193 p. 0231121989 TN900 553.6 2001-28227

History, chemistry, physics, economics, anthropology, technology, language and linguistics, art history, advertising taglines, and culinary arts are all explored in this wonderful, multicultural Renaissance approach to the subject of salt. From developments in the technology of salt production and the impact on the wealth and social progress of the areas that possessed this capability, Laszlo's analysis shows that salt is inextricably intertwined into the basic fabric of our lives, culture, and language.

8.042
Smithells metals reference book, ed. by E. A. Brandes and G.B. Brook. 7th ed. Oxford; Boston: Butterworth-Heinemann, 1998. Paperback reprint with corrections, 1999. 1 v., various pagings. 0750636246 pbk TN671 669 97-27401

New edition of Colin James Smithells's longstanding reference; 6th ed. in 1983; first published in London in 1949. Extensive references and good summaries. Revisions in the newer editions included corrosion, friction and wear, metallography, and soldering and brazing.

Perfumes

8.043
The chemistry of fragrances, compiled by David Pybus and Charles Sell. Cambridge, UK: Royal Society of Chemistry, 1999. 276 p. (RSC paperbacks) 0854045287 TP983 668.54

Presents the latest chemistry, techniques, and tools applied to fragrance creation. Discussions range through the history of perfumes to the structure of the perfume industry today. Treats consumer research, toxicology, and use of electronic odor-detection technology.

Plastics

Handbook and Encyclopedia
8.044
Handbook of plastics, elastomers, and composites. 3rd ed. Charles A. Harper, ed. in chief. NY: McGraw-Hill, 1996. various pagings 007026693x TP1130 668.4 96-20787

Data cover fundamentals of plastics and elastomers as well as applications. Glossary and index are provided.

History and Development
8.045
Fenichell, Stephen. *Plastic: the making of a synthetic century*. NY: HarperCollins, 1996. 356 p. 0887307329 TP1116 668.4 96-364

A relatively evenhanded treatment of the plastics industry, full of anecdotes about the history of polymers and their impact on humanity. Plastics have had both good and bad effects, and this book offers up both to the reader. A social commentary rather than a chemistry book, with history, social context, and opinion.

8.046
Meikle, Jeffrey L. *American plastic: a cultural history*. New Brunswick, NJ: Rutgers University Press, 1995. 403 p. 081352234x; 0813522358 pbk TP1117 303.48 95-15187

Chapters are dedicated to celluloid, Bakelite, nylon, and other major classes of synthetic polymers and plastics. Intellectually imposing but written in a nonscientific and literary style. A definitive resource for scholarly research into the history and sociology of plastics.

8.047

The Plastics age: from Bakelite to beanbags and beyond, ed. by Penny Sparke. Woodstock, NY: Overlook Press, 1993. 160 p. 087951471x; 0879514884 pbk TP1122 668.4 92-35039

In three parts: Plastics pre-history, 1860-1914; Plastics and modernity, 1915-1960; and Plastics and post-modernity, 1961-1990. Chapters are by authorities on various aspects of the history and applications of plastics. Beautifully illustrated; more than 150 photographs.

Properties

8.048

Brydson, J.A. *Plastics materials.* 7th ed. Boston: Butterworth-Heinemann, 1999. 920 p. 0750641320 TP1120 668.4 99-30623

Explains the preparation of plastic materials and how properties are related to chemical structures. For each plastic material, preparation, structure, and properties are provided.

8.049

Gruenwald, Geza. *Plastics: how structure determines properties.* Munich; NY: Hanser Publishers, 1993. 357 p. (SPE books) 3446165207 (Hanser); 0195209583 (Oxford) TA455 620.1 92-53928

Discusses bonding, architecture, and additives in polymers that affect processing, mechanical properties, and chemical properties in the resulting product. Very informative discussions of crystallization processes, annealing, and effects on remelting on plastic properties.

Environmental Considerations

8.050

Stevens, Eugene S. *Green plastics: an introduction to the new science of biodegradable plastics.* Princeton: Princeton University Press, 2002. 238 p. 069104967x TP1180.B55 668.4 2001-36257

Introduces plastics, i.e., polymers; these materials are discussed with respect to chemical makeup, use in society, their degradation properties, and how that is involved in plastics' effects on the environment. Also focuses on what are now being called "bioplastics"; these are biodegradable plastics whose components are derived entirely or almost entirely from renewable raw materials. Bioplastics found in history are examined as are their reappearance in today's world, along with the future prospects and the factors affecting the materials' growth. Useful appendixes; glossary; list of books for suggested reading; directions on how to "make your own" bioplastics.

8.051

Wallace, Deborah. *In the mouth of the dragon.* Garden City Park, NY: Avery Pub. Group, 1990. 230 p. 0895294400 TH9446 363.17 89-18239

Treats the fire-related decomposition of plastics and the dangers of fires fueled by plastics. Six chapters describe major fires involving plastics that caused multiple deaths. Other chapters discuss chemical properties of polymers and laboratory tests of polymer degradation at high temperatures.

8.052

Wolf, Nancy, and Ellen Feldman. *Plastics: America's packaging dilemma.* Washington, DC: Island Press, 1991. 131 p. (Island Press critical issues series, 3) 1559630639; 1559630620 pbk TD798 363.72 90-43801

Designed for lay readers who may be tackling issues of recycling, the book focuses on disposal options (recycling, incineration, landfilling), and current controversies (definitions of degradability, enforced coding of resin content, problems with multi-layered materials); however, there is a definite bias against chemicals.

Petrochemicals

8.053

Berkowitz, Norbert. *Fossil hydrocarbons: chemistry and technology.* San Diego: Academic Press, 1997. 351 p. 012091090x TP343 553.2 97-23439

Overviews heavy hydrocarbons and their potential uses in the industrial, utility, and transportation sectors. Integrates material from standard professional handbooks, with discussion and examples. Thoroughly explains origins, geochemistry, and classification of heavy hydrocarbons, followed by chemical and physical properties of various hydrocarbons. Illustrations and graphs explain and illustrate properties. Later chapters highlight key technologies and chemical processes used to enhance the value of hydrocarbons, via purification, chemical modification (e.g., thermal cracking), and physical conversion.

Rubber and Composite Materials

8.054

White, James Lindsay. *Rubber processing: technology, materials, principles.* Munich; NY; Hanser Publishers; Cincinnati: Hanser/Gardner Publications, 1995. 586 p. 1569901651 TS1890 678 93-48549

Offers an overview of rubber and polymer descriptions, including general rubber and polymer materials, and mathematical models of such topics as kinematics of flow, viscoelastic fluid models, Newtonian fluid flow, and flow in calendering.

Dyes and Colorants

8.055

Balfour-Paul, Jenny. *Indigo.* London: British Museum Press, 1998. 264 p. 0714117765 TP923 667 99-223017

Cultural history—social, political, and economic—of dyed fibers over time, and the special history of indigo dye, from discovery to chemical synthesis.

8.056

Christie, R.M. *Colour chemistry.* Cambridge: Royal Society of Chemistry, 2001. 205 p. (RSC paperbacks) 0854045732 TP910 667.2

Provides up-to-date insight into the chemistry behind the color of dyes and pigments. Discusses the history, structure, and synthesis of dyestuffs, and applications to textiles.

Paper Chemistry

8.057

Paper chemistry, ed. by J.C. Roberts. Glasgow: Blackie: NY: Chapman and Hall, 1991. 234 p. 0216929091; 0412025116 TS1120 676 90-27806

Treats the applications of paper chemistry, accessibility of cellulose, electrokinetics of papermaking systems, retention aids, dry strength additives, wet strength in paper, sizing of paper with rosin and alum at acid pHs, neutral and alkaline sizing, dyes and fluorescent whitening agents for paper, physical and chemical aspects of fillers in papermaking, measurement and control, and practical applications.

Pyrotechnics and Explosives

8.058

Russell, Michael S. *The chemistry of fireworks.* Cambridge, UK: Royal Society of Chemistry, 2000. 117 p. (RSC paperbacks) 0854045988 TP300 662.1

Glossary of British terms precedes a dozen chapters, beginning with a historical introduction up through Greek technology, Chinese developments, Roger Bacon's contributions, black powder, rockets, mines and shells, fountains, sparklers, bangers, Roman candles, and gerbs and wheels. Ends with special effects, which treats fuses, lances, setpieces and devices, and other technologies. Two final chapters address safety and legislation.

Chapter 9

Polymer Chemistry

This installment of the trek through the literature of chemistry will look at the world of polymers in a very general way. Polymeric substances have already been examined in the light of materials science; here, polymer chemistry information sources of all types are examined. This listing is by no means everything there is to know about polymers; there doubtless will be more publications by the time this book is published. As the types of polymers and their applications multiply, so will the literature of the field expand. (Plastics, if the reader remembers, were included in the chapter on industrial chemistry, because they are a product, rather than a chemical entity.)

GUIDE TO THE LITERATURE

9.001

Information sources in polymers and plastics, ed. by R.T. Adkins. London; NY: Bowker-Saur, 1989. 313 p. (Guides to information sources) 040802027x Z5524 016.5477 89-37691

Chapters treat serial literature: primary journals, abstracts; books, encyclopedias, and reviews; patents and trademarks; standards; trade literature, theses, conferences; online databases; polymer structures and nomenclature; additives and catalysts; properties of plastics and polymers; business information; fibers; rubber; coatings and adhesives; new developments in polymers: electronics and electrolytes; high-performance plastics; Europe (excluding the UK, Scandinavia, and the CMEA countries); the plastics industry in the countries of the CMEA; the Americas: North and South; the Pacific basin and India; translations.

GENERAL RESOURCES

9.002

Seymour, Raymond Benedict. *Seymour/Carraher's polymer chemistry.* 6th ed., rev. and expanded, Charles E. Carraher

Jr. NY: M. Dekker, 2003. 913 p. (Undergraduate chemistry, 16) 0824708067 QD381 547

Introduces polymer science; polymer structure; molecular weight; testing and spectrometric characterization; rheology and physical tests; naturally occurring polymers; step-reaction polymerization (polycondensation); ionic chain-reaction and complex coordinative polymerization (addition); free-radical chain polymerization (addition); copolymerization; inorganic-organic polymers; inorganic polymers; fillers and reinforcements; plasticizers, stabilizers, flame retardants, and other additives; reactions; synthesis of reactants and intermediates; and polymer technology.

9.003

Tonelli, Alan E., with Mohan Srinivasarao. *Polymers from the inside out: an introduction to macromolecules.* NY: Wiley-Interscience, 2001. 249 p. 0471381381 QD381 547 00-47990

Introduces synthesis and properties, step and chain growth polymer synthesis, microstructure and conformations of polymers, solution and bulk polymer properties, and biopolymers. Numerous illustrations. Includes some nice experiments that can be used as demonstrations, along with descriptions of common polymer science characterization methods.

NOMENCLATURE

9.004

List of standard abbreviations (symbols) for synthetic polymers and polymer materials, 1974. International Union of Pure and Applied Chemistry, Commission on Macromolecular Nomenclature. Oxford; NY: Pergamon, 1978. pp. 475-491 in *Pure & Applied Chemistry.* 0080223710 QD7

Published in *Pure & Applied Chemistry,* vol. 40, no. 3, 1974, for the first time.

ENCYCLOPEDIAS

9.005

Cheremisinoff, Nicholas P. *Condensed encyclopedia of polymer engineering terms.* Boston: Butterworth-Heinemann, 2001. 362 p. 0750672102 TP1087 668.9 00-68904

Cheremisinoff, author/editor of a large number of reference works in engineering and chemistry, has produced a readable and informative guide providing both theoretical and practical information. The approximately 400 entries, ranging from a sentence to a few pages, are often supplemented by helpful diagrams, schematics, and charts. Emphasizes properties, processing, and applications of commercially available polymers. Tables and discussions include materials from a number of manufacturers.

9.006

Concise encyclopedia of polymer science and engineering. Jacqueline I. Kroschwitz, executive ed. NY: Wiley, 1990. 1,341 p. (A Wiley-Interscience publication) 0471512532 TP1087 668.9 89-70674

Said to contain all subjects covered in the 17 main volumes and supplement and index volumes of the *Encyclopedia of Polymer Science and Engineering*, 2nd ed. Treats polymeric materials, both natural and synthetic; polymer properties (molecular, chemical, physical, electrical, mechanical, thermal, and biological), morphology, compatibility, and stability; synthesis and reactions; characterization and analytical methods; physical processes, engineering; polymer processing; product fabrication; test methods, uses in adhesives, coatings, films, fibers, elastomers, plastics composites, and occurrence in natural materials; historical perspective; and economics.

9.007

Concise polymeric materials encyclopedia; ed. in chief, Joseph C. Salamone. Boca Raton, FL: CRC Press, 1999. 1,706 p. 084932226x TP1110 668.9 98-6146

More than 1,000 articles in this work are derived from *Polymeric Materials Encyclopedia* (12v., cited below). Topics include development of new catalysts to modify polymers and synthesize new ones, and synthesis, preparation, properties, and applications of new compounds. Particular attention is given to biologically oriented polymers and to those with health-care applications. The brief review articles are organized in sections and are cross-referenced to related subjects. Structural diagrams and figures; extensive

bibliography of patents, well-known journals, and other sources in polymer science. Materials are indexed in both polymeric and monomeric forms. Available online via Dialog.

9.008

Encyclopedia of polymer science and engineering. 2nd ed. Editorial board, Herman F. Mark ... [et al.]; ed. in chief, Jacqueline I. Kroschwitz. NY: Wiley, 1985-<1990 > v. <1-17 >; 1 supplement + <6 >. (A Wiley-Interscience publication) 0471895407 (v. 1); 0471809489 TP1087 668.9 84-19713

Volumes contain authoritative, lengthy, signed articles on selected topics in polymer science and related engineering processes. Articles include brief introduction, composition and properties of compounds and products, manufacturing processes, applications, and economic aspects. Separate index volume contains the table of contents for all volumes. Rev. ed. of Encyclopedia of polymer science and technology, 1964- .

9.009

Encyclopedia of polymer science and engineering; Pt. 1, ed. by Jacqueline I. Kroschwitz. 3rd ed. Hoboken, NJ: Wiley-Interscience, 2003- . 4 v. 0471288241 (Pt. 1); 0471809489 TP1110 668.9 2003-41107

Pt. 1: v. 1: A to coatings; v. 2: Coextrusion to hyperbranched polymers; v. 3: Injection molding to polysulfides; v. 4: Polysulfones to weathering. Set to be complete by April 2004, in 12 v. (ISBN 0471275077). Also known as the *Mark Encyclopedia of Polymer Science and Technology*, originally by Herman F. Mark.

9.010

Polymeric materials encyclopedia; ed. in chief, Joseph C. Salamone. Boca Raton, FL: CRC Press, 1996. 12 v.; includes CD-ROM version 084932470x (set); 0849326516 (CD-ROM) TP1110 668.9 96-12181; 96-12182 (CD-ROM)

A comprehensive and detailed work on the chemical and chemical engineering aspects of polymer synthesis and characterization. Mostly synthetic polymers are covered, but natural polymers are included. Alphabetically arranged long articles cover various general and specific chemical classes of polymers, with discussion of particular applications and methods interspersed among the articles. Current to 1995.

HANDBOOKS

9.011

CRC materials science and engineering handbook.
2nd ed. Editor, James F. Shackelford; associate editor,
William Alexander; assistant editor, Jun S. Park. Boca
Raton, FL: CRC Press, 1994. 1,545 p. 0849342503 TA403.4
620.1 94-163322

A useful addition to the reference tools of engineers and
materials scientists. There is good coverage of both organic
and inorganic materials, as well as metals, glasses,
semiconductors, and polymers. Mechanical properties,
corrosion data, and thermodynamic properties are included.

9.012

Polymer handbook. 4th ed. Editors, J. Brandrup, E.H.
Immergut, and E.A. Grulke; associate editors, A. Abe, D.R.
Bloch. NY: Wiley-Interscience, 1999. 1 v., various pagings
(A Wiley-Interscience publication) 0471166286 QD388 547
98-37261

Brings together data for theoretical and experimental
polymer research. Includes fundamental, validated property
data; new developments since 1989; improved nomenclature
information; polymerization and depolymerization; physical
properties of monomers and solvent; physical data on
oligomers; physical constants of some important polymers;
solid state properties; solution properties; and abbreviations
of polymer names. Numerous tables on polymer synthesis,
degradation, kinetics, thermodynamics, analysis and
characterization, solid state and solution data, and solvents
and oligomers.

DICTIONARY

9.013

Alger, Mark S.M. *Polymer science dictionary*. London;
NY: Elsevier Applied Science, 1989. 532 p. 1851662200
QD380.3 547.7 88-11034

Entries discuss polymer chemistry, polymer physics, and
polymer engineering. Classes of polymers, polymer-forming
reactions, and important polymers are included, as well as
applied topics.

HISTORY

9.014

Furukawa, Yasu. *Inventing polymer science: Staudinger,
Carothers, and the emergence of macromolecular
chemistry*. Philadelphia: University of Pennsylvania Press,
1998. 310 p. (The Chemical sciences in society series)
0812233360 QD381 547 97-36390

A fascinating account of the development of polymer
science primarily from the view of two of the main
protagonists in its inception. Well-based in both science and
historical research, with a comprehensive, detailed account
of the resolution of a major controversy in science.

TREATISES AND COMPREHENSIVE WORKS

9.015

Billmeyer, Fred W., Jr. *Textbook of polymer science*.
3rd ed. NY: Wiley, 1984. 578 p. (A Wiley-Interscience
publication) 0471031968 QD381 668.9 83-19870

A basic text in the field, this work contains information on
all types of polymers.

9.016

Boyd, Richard H., and Paul J. Phillips. *The science of polymer molecules: an introduction concerning the synthesis, structure and properties of the individual molecules that constitute
polymeric materials*. Cambridge; NY: Cambridge University
Press, 1993. 410 p. (Cambridge solid state science series)
0521320763 QD281 547.7 94-153475

This introductory book focuses on basic chemistry of
polymers; kinetics and mechanisms are introduced. Three-
dimensional structures are examined. Statistical behavior of
polymer chains, rubber elasticity, and solutions are
discussed.

9.017

***Comprehensive polymer science: the synthesis,
characterization, reactions & applications of
polymers***. Chairman of the editorial board, Sir Geoffrey
Allen; deputy chairman of the editorial board, John C.
Bevington. Oxford, England; NY: Pergamon Press, 1989. 7
v. First supplement, 1992; second supplement, 1996.
0080325157 (set) QD381 547.7 88-25548

V. 1: Polymer characterization; volume editors, Colin Booth

& Colin Price. V. 2: Polymer properties; volume editors, Colin Booth & Collin Price. V. 3: Chain polymerization I; volume editors, Geoffrey C. Eastmond [et al. V. 4: Chain polymerization II; volume editors, Geoffrey C. Eastmond [et al.]. V. 5: Step polymerization; volume editors, Geoffrey C. Eastmond [et al.]. V. 6: Polymer reactions; volume editors, Geoffrey C. Eastmond [et al.]. V. 7: Specialty polymers & polymer processing; volume editor, Sundar L. Aggarwal. Supplement volumes bring the set up to date.

9.018

Hall, Christopher. *Polymer materials: an introduction for technologists and scientists.* 2nd ed. NY: Wiley, 1989. 243 p. (A Halsted Press book) 0470210923 TA455 620.1 88-5520

The material properties of polymers are discussed. First published in 1989 by Macmillan Education Ltd., London.

9.019

Misra, Gauri Shankar. *Introductory polymer chemistry.* NY: J. Wiley & Sons, 1993. 253 p. 0470217200; 8122404715 (Wiley Eastern Ltd.) QD381 547.7 90-21092

Introduces polymers to two-year students. Includes physical, mathematical, and chemical bases of polymerization mechanisms. Explores free-radical, cationic and anionic, condensation, and emulsion polymerization, as well as copolymers, additives, polymer solubility, viscosity, cross-linking, and other pertinent factors.

9.020

Munk, Petr, and Tejraj M. Aminabhavi. *Introduction to macromolecular science.* 2nd ed. NY: Wiley, 2002. 609 p. 0471417165 QD381 547 2001-45583

This new edition (1st ed., 1989) is for beginning students and includes five chapters on structure, synthesis, solutions, bulk polymers, and technology, which range in length from 36 to 233 pages each. Also included is an epilogue (i.e., appendix) listing textbooks, monographs, encyclopedias, handbooks, and journals devoted to macromolecules (i.e., polymers). The chapter on solutions is the most extensive and includes a wide range of topics, mostly at a qualitative or descriptive level.

9.021

Painter, Paul C., and Michael M. Coleman. *Fundamentals of polymer science: an introductory text.* 2nd ed. Lancaster,

PA: Technomic Pub. Co., 1997. 478 p. 1566765595 QD381 547.7 97-60515

This second edition (1st ed., 1994) includes study questions and suggestions for further reading, and is an expansive overview with polymer synthesis, various types of polymerization and copolymerization, and polymer theories and statistics. Polymer characterization focuses mainly on spectroscopic methods. Physics and physical polymer chemistry includes structure and morphology, crystallization and melting, solution thermodynamics, polymer blends, and molecular weight and branching. Mechanical and rheological properties are also treated. History of some significant developments in polymer science is also included.

9.022

Ravve, A. *Principles of polymer chemistry.* NY: Plenum Press, 1995. 496 p. 0306448734 QD381 547.7 95-7219

Eight chapters treat ring-opening polymerization, natural polymers, polymer modification, commercially significant chain growth polymers and their industrial preparations, kinetics of thermoplastic and thermoset polymerization, and to a lesser degree, emulsion and Ziegler-Natta polymerization and stereochemistry of polymerization. Accompanying diskette.

9.023

Stevens, Malcolm P. *Polymer chemistry: an introduction.* 3rd ed. NY: Oxford University Press, 1999. 551 p. 0195124448 QD381 547 98-23083

Introduces principles of polymerization, polymer properties, polymer characterization, polymerization methods for vinyl polymers (free radical, ionic, Ziegler-Natta), reactions on polymers, step-growth and ring-opening polymerizations, high performance thermoplastics, heterocyclics, thermoset polymers, inorganic polymers, and natural polymers. Appendixes list polymer abbreviations, polymer reference books and journals, and resources for experiments for polymer laboratory classes.

TECHNIQUES IN POLYMER CHEMISTRY

9.024

Akelah, A., and A. Moet. *Functionalized polymers and their applications.* NY: Chapman and Hall, 1990. 354 p. 041230290x TP1087 620.1 90-223124

A comprehensive survey of the literature and an extensive review of current research in this field. Nine chapters average 250-400 references each, current through 1986; material ranges from preparative chemistry and properties through chemical applications, biological applications, and other types of applied polymer science. A second part reviews polymeric reagents and catalysts, separations and functionalized polymers, and organic synthesis on polymeric carriers. Other parts discuss controlled release formulations and polymeric support for active groups, conductive polymers, photoresists, liquid crystals, and other novel materials such as wood preservatives and polymeric fertilizers. Good index.

9.025

Collins, Edward A., Jan Bares, and Fred W. Billmeyer, Jr. *Experiments in polymer science.* NY: Wiley, 1973. 530 p. (A Wiley-Interscience publication) 0471165840; 0471165859 pbk QD385 547 73-650

This volume is a companion to Billmeyer's *Textbook*, and is suitable for advanced study.

9.026

Sandler, Stanley R., and Wolf Karo. *Sourcebook of advanced organic laboratory preparations.* San Diego: Academic Press, 1992. 332 p. 0126185069; 0126185506; 0126185115 QD262 547.2 92-13631

This book is a compilation of selected lab preparations from Sandler's previous books, one of which was *Polymer Synthesis*, 1980. Procedures are from literature sources. Chapters include a brief introduction to the general preparation of the functional group, with a few representative procedures. Syntheses for several high polymers are included. Index of name reactions used in the text; good references.

9.027

Seymour, Raymond Benedict. *Engineering polymer sourcebook.* NY: McGraw-Hill, 1990. 300 p. 0070563608 TP156 668.9 89-35066

Each of 20 chapters treats a different polymer or class of polymers, emphasizing engineering properties related to design, processing, fabrication, and application in construction.

9.028

Sorenson, Wayne R., and Tod W. Campbell. *Preparative methods of polymer chemistry.* 2nd ed. NY:

Interscience Publishers, 1968. 504 p. 0471813796 QD281 547 67-29543

A laboratory manual as well as a reference work; includes references, figures, photographs, and tables.

CHARACTERIZATION OF POLYMERS

General Methods

9.029

Campbell, D., and J.R. White. *Polymer characterization.* Chapman and Hall, 1989. 350 p. 0412271605; 0412271702 pbk QD381.8 547.7 89-34186

Discusses traditional wet chemical and thermal techniques for characterizing polymers, but emphasizes techniques for probing structures. Tools mentioned include UV, IR, Raman, NMR, ESR, X-ray diffraction, and electron microscopy.

9.030

Characterization of polymers, ed. by Ho-Ming Tong [et al.]. Boston: Butterworth-Heinemann; Greenwich: Manning, 1994. 319 p. (Materials characterization series) 0750692871 TA455 620.1 93-1909

Chapter 1 treats polymer structures and synthesis; chapter 2 summarizes polymer fabrication techniques, including polymer insulators for the space shuttle; Langmuir-Blodgett films, and fiber drawing. Other chapters discuss chemical composition using X-ray photoelectron spectroscopy, FTIR, Raman, and mass spectroscopy; microscopic characterization of films; light scattering characterization of interfaces; wetting, spreading, adhesion, and other interfacial properties of polymers; surface modifications that enhance wettability, weldability, printability, resistance to craze, and adhesion; polymer interfaces with metals and ceramics; polymer-polymer interfaces; tribology; and some ASTM testing methods.

9.031

Polymer characterisation, ed. by B.J. Hunt and M.I. James. London; NY: Blackie Academic & Professional, 1993. 362 p. 0751400823 QD139 547.7 93-116908

Chapters treat separation and analysis of additives, NMR characterization, vibrational spectroscopy, molecular weight determination, chromatographic determination, thermal analysis, small-angle neutron scattering, neutron reflectometry, mechanical and rheological testing, microscopy, and characterization by XPS and SIMS.

Spectroscopy

9.032

Bovey, Frank A., and Peter A. Mirau. *NMR of polymers*. San Diego: Academic Press, 1996. 459 p. 0121197654 QP519.9 574.19 96-282240

A comprehensive overview of NMR applications specifically relating to polymer science. Five chapters thoroughly review NMR fundamentals and experimental techniques; characterization of polymer microstructure including stereochemistry, optically active side chains, branching, and copolymer sequences: multinuclear and two-dimensional NMR methods; solid-state NMR, with chain conformation studies, multiphase polymers, and polymer blends; and molecular dynamics of polymers in both solution and solid state.

9.033

Bower, D.I., and W.F. Maddams. *The vibrational spectroscopy of polymers*. Cambridge; NY: Cambridge University Press, 1989. 326 p. 0521246334; 0521421950 pbk QD381.8 547.8 88-2956

Provides information on theory and practice, including basic infrared and Raman, application of symmetry elements, vibrational modes, and assignments; characterizes and interprets spectra for major classes of polymer spectra; provides group frequencies; introduces quantitative analysis and some instrumentation; and discusses microstructures and polymerization processes.

9.034

The infrared spectra atlas of monomers and polymers. Richard A. Nyquist, ed. Philadelphia, PA: Sadtler Research Laboratories, 1980. 810 p. 0845600648 QC463 547.3 80-53761

Spectra of 2,000 monomers and polymers selected from the Sadtler collection of spectra, organized by polymer class; brief description of group frequency characteristics, names, manufacturers, and analysis type are given.

9.035

Koenig, Jack L. *Spectroscopy of polymers*, 2nd ed.

Amsterdam; NY: Elsevier, 1999. 491 p. 0444100318 QD139.P6 547 99-37577

9.036

Mass spectrometry of polymers, ed. by Giorgio Montaudo and Robert P. Lattimer. Boca Raton, FL: CRC Press, 2002. 584 p. 0849331277 QD139.P6 547 2001-37684

Mass spectrometry (MS) together with NMR are the indispensable tools of chemistry and physics for characterizing polymer properties. Over the past half-century, polymer MS has now emerged as a field in its own right for high molecular weight, dispersity, endgroups, impurities, additives, copolymers, blends and mixtures, unnatural and natural biopolymers. This skillful rendering provides two of the best introductory chapters, by way of general introduction to MS and summary of MS characterization methods. Altogether, there are 11 chapters, and more than 1,000 references (current to 2001).

9.037

Nyquist, Richard A. *The infrared spectra building blocks of polymers*. Philadelphia, PA: Sadtler Research Laboratories, 1989. 268 p. 0845601512 QC462.85 547.3 88-60355

Presents theory and spectra-structure correlations. Contains some 1,000 IR spectra of about 700 compounds. Displays spectra-structure correlations for carbon-carbon double and/or triple bonded hydrocarbons as well as compounds with functional groups. Compounds include (besides spectra) molecular formulas, molecular weights, CASRNs, melting and boiling points, source of spectra, and techniques used in sample preparation.

9.038

The Sadtler guide to carbon-13 NMR spectra of polymers and resins. Philadelphia, PA: Sadtler, 1988. 353, 7, 7 p. 0845601504 QC762 547.7046 88-60354

Contains more than 350 carbon-13 NMR spectra of polymers, copolymers, and resins. Polymer types are those from monomers including ethylene, propylene, isobutylene, butene, butadiene, isoprene, chloroprene, acrylonitrile, vinyl chloride, vinyl alcohol, vinyl acetate, vinyl pyrrolidone, styrene, urethane, sulfide, and carbonate; and other polyesters, nylon, and silicones. Resin types include those from hydrocarbon, epoxies, coumarone-indene, phenolics, acrylics, rosins, and rosin esters.

9.039

Simons, William W., and M. Zanger. *The Sadtler guide to the NMR spectra of polymers.* Philadelphia, PA: Sadtler Research Laboratories, 1973. 298 p. 0845600028 QD139 547 73-90432

Compounds are selected from the Sadtler Standard NMR Collection. Analysis of polymers by proton NMR spectroscopy. Divided into 12 classifications, providing chemical shift, coupling constant, and other significant features.

Miscellaneous Methods

9.040

Hatakeyama, T., and F.X. Quinn. *Thermal analysis: fundamentals and applications to polymer science.* Chichester; NY: Wiley, 1994. 158 p. 0471983624 QD79 543 98-49129

A practical guide and general introduction to thermal analysis techniques, focusing on which techniques should be avoided and which are best for sample preparation.

9.041

Woodward, Arthur E. *Understanding polymer morphology.* Munich; NY: Hanser Publishers; Cincinnati: Hanser/Gardner Publications, 1995. 130 p. (Hanser understanding books) 1569901414 TP1087 668.9 94-37332

Introduces the basics of morphological investigations followed by detailed discussions of polymer morphology. Observed morphologies are correlated with given polymer types, such as semicrystalline, liquid crystals, block copolymers, and blends. Also discussed are morphologies resulting from processing, deformation, and failure; and effects of morphology on mechanical properties.

Mechanical and Chemical Behavior of Polymers

9.042

Courtney, Thomas H. *Mechanical behavior of materials.* 2nd ed. NY: McGraw-Hill, 2000. 0070285942 TA405 620.1 99-39661

The relationships between macro and micro properties, together with fundamental concepts of crystal structure and bonding, are used to develop a quantitative description of structural relationships, not only in polymers but in metals, ceramics, and composites.

9.043

Handbook of polymer degradation, ed. by S. Halim Hamid, Mohamed B. Amin, and Ali G. Maadhah. NY: M. Dekker, 1992. 649 p. (Environmental science and pollution control series, 2) 0824786718 QD381.9 620.1 92-10169

Thoroughly discusses photodegradation, thermodegradation, biodegradation, and the effects of environmental influences such as acid rain. Complete descriptions are provided for techniques used to determine the rate of degradation under various conditions. Chemical mechanisms are clearly described, and various instrumental methods for measuring degradation are discussed.

Physical and Chemical Properties of Polymers

9.044

Physical properties of polymers handbook, ed. by James E. Mark. Woodbury, NY: AIP Press, 1996. 723 p. (AIP series in polymers and complex materials) 1563962950; 1563965984 (CD-ROM); 1563965992 (set) TA455 620.1 95-50256

Topical areas treated include theory (rotational isomeric state models and results), spectroscopy (including neutron and X-ray scattering), mechanical properties (including adhesives, composites, and gels), crystallinity and morphology, and electro-optical and magnetic properties (including electrical conductivity, electroluminescence, nonlinear optics, and piezo-, pyro-, and ferroelectric properties). Discussions include methodology and test methods.

9.045

Ragone, David V. *Thermodynamics of materials.* NY: Wiley, 1995. 2 v.: 311 p., 242 p. (MIT series in materials science and engineering) 0471308854 (v. 1); 0471308862 (v. 2) TA418.52 536 94-25647

Treats energy and transformations in various classes of materials, such as polymers, ceramics, and semiconductors. Statistical thermodynamics is used as a background for understanding differences in the classes of materials.

9.046

Strobl, Gert R. *The physics of polymers: concepts for understanding their structures and behavior.* 2nd corr. ed. Berlin; NY: Springer-Verlag, 1997. 439 p. 3540632034 pbk QC173.4 530.4 97-28362

Chapters treat constitution and architecture of polymer chains, single-chain conformations, liquid equilibrium states, metastable partially crystalline states, mechanical and dielectric response, microscopic dynamical models, nonlinear mechanical behavior, and yield processes and fracture. A final chapter discusses scattering experiments on polymer microstructures.

9.047

Zoller, Paul, and David J. Walsh. *Standard pressure-volume-temperature data for polymers.* Lancaster, PA: Technomic Pub. Co., 1995. 412 p. 1566763282 QD381.9.T54 95-60846

PVT data are presented in graph and table form for some 180 polymers. Contents: introduction, hydrocarbons, hydrocarbon polymers, ethylene polymers, styrenics, acrylics, polyacrylonitrile and copolymers, other C-C main chain polymers, polyethers, polyamides, polyesters, various main chain aromatics, blends, and miscellaneous.

MULTICOMPONENT SYSTEMS

9.048

Sperling, L.H. *Polymeric multicomponent materials: an introduction.* NY: Wiley, 1997. 397 p. (A Wiley-Interscience publication) 0471041386 QD381 620.1 97-6509

Reviews essential terms, concepts, theories, and experimental facts and procedures concerning polymer-polymer and polymer-nonpolymer combinations. Also covers interpenetrating polymer networks, graft and block copolymers, structural composites, coatings, carbon black-reinforced elastomers, and fiber-reinforced plastics.

Chapter 10

Biological Chemistry

This chapter covers the literature of the basic areas of biological chemistry in much the same manner as was chemistry itself. This part of chemistry includes natural products (chemicals found in humans, animals, insects, plants, and fungi, for instance), pharmaceuticals, and the basic studies of just what constitutes life. A final portion of the chapter treats special topics: proteins, medicinal chemistry, lipids, food and nutrition, toxicology, and some plant products, all as they relate to biochemistry.

The Library of Congress areas represented in biological chemistry areas, besides the QD sections, are QP, physiology, and the QP501-801 sections, representing animal biochemistry.

GUIDES TO THE LITERATURE

10.001

Schmidt, Diane, Elisabeth B. Davis, and Pamela F. Jacobs. *Using the biological literature: a practical guide.* 3rd ed., rev. and expanded. NY: Marcel Dekker, 2002. 474 p. (Books in library and information science, 60) 0824706676 QH303.6 570 2001-58392

Offers the history and characteristics of biological literature, reviews various subject headings and classification systems, and then discusses all the major areas of biology with respect to their literatures.

10.002

Information sources in the life sciences, ed. by H.V. Wyatt. 4th ed. London; New Providence, NJ: Bowker-Saur, 1997. 264 p. (Guides to information sources) 1857390709 QH303.6 016.57 96-29868

Includes bibliographies and index to book titles, subjects, and major formats. Discusses current awareness; computerized searching; abstracts, indexes, and bibliographies; major research databases; guides to the literature; biochemical sciences; microbiology;

biotechnology; genetics; zoology; ecology; botany; and history of biology.

ABSTRACTING AND INDEXING SERVICES

10.003

BasicBIOSIS [electronic resource]. Philadelphia: BIOSIS; Dublin, OH: OCLC, Inc. Updated monthly; coverage is last four years of publication. QH 301

Available through OCLC's FirstSearch service.

10.004

Biological abstracts. Philadelphia: BioSciences Information Service of Biological Abstracts, v. 1, Dec. 1926- . Monthly (irregular), 1926- ; semimonthly, Dec. 15, 1975- . ISSN 0006-3169 QH301 570.5 31-13663

Volumes for 1939-1962 published in sections A, General biology; B, Experimental animal biology, 1939-47; B, Basic medical sciences, 1948-62; C, Microbiology, immunology and parasitology; D, Plant sciences; E, Animal sciences; F-J added after 1939. Also available from 1990- in CD-ROM as *Biological Abstracts on Compact Disc*, and online as part of *BIOSIS Previews.*

10.005

Biological abstracts/RRM. (Biological abstracts, reports, reviews, meetings). Philadelphia: Biosciences Information Service, v. 18, 1980- . Semimonthly, <1982- > ISSN 0192-6985 Z5321 016.574 80-642178

Also available as *Biological Abstracts/RRM* on compact disc since 1991. A supplement to *Biological Abstracts* that offers reports, reviews, meetings, and books; there are no abstracts but keywords are listed. Author, biosystematic, generic, and subject indexes are provided.

10.006

BIOSIS previews on compact disc [electronic resource]. Norwood, MA: SilverPlatter International.

Monthly CD-ROMs, archival retrospective discs. QH301 sn99-48840

Supplies coverage of international life science journal and meeting literature found in *Biological Abstracts* and *Biological Abstracts/RRM*. Also available on the Internet via the SilverPlatter server.

10.007

PubMed. Online service that serves as access for MEDLINE, at http://ncbi.nlm.nih.gov/entrez/.

PubMed is a service of the National Library of Medicine and includes more than 14 million citations for biomedical articles reaching back to the 1950s. These citations are from MEDLINE and additional life science journals. *PubMed* includes links to many sites that offer full-text articles and other related sources. The service is free and offers abstracts.

ENCYCLOPEDIAS

10.008

Brockhaus ABC Biochemie. *Concise encyclopedia biochemistry and molecular biology*. English language ed., 3rd ed. [Translated into English], rev. and expanded by Thomas A. Scott and E. Ian Mercer. Berlin; NY: Walter de Gruyter, 1997. 737 p. 3110145359 QD415 572 9647538

(Rev. ed. of *Concise Encyclopedia Biochemistry*. English language ed., 2nd ed., 1988) Resembles the publisher's *Concise Encyclopedia Biology*. The material is arranged alphabetically under broad headings; article length varies from one paragraph to five pages, with a level of detail appropriate to undergraduates as well as faculty. Line drawings; few references; idiomatic translation from German.

10.009

The encyclopedia of bioprocess technology: fermentation, biocatalysis, and bioseparation, [ed. by] Michael C. Flickinger and Stephen W. Drew. NY: Wiley, 1999. 5 v., 2,756 p. (Wiley biotechnology encyclopedias; A Wiley-Interscience publication) 0471138223 (set) TP248.3 660.6 99-11576

Some 300 well-chosen articles, signed, well written and organized, with keywords and outlines, and good bibliographies. Good index; no glossary.

10.010

The Encyclopedia of molecular biology; editor in chief

Sir John Kendrew; exec. ed., Eleanor Lawrence. Oxford; Cambridge, MA: Blackwell Science, 1994. 1,165 p. 0632021829 QH506 574.8 94-22951

Treats structural biology, molecular genetics, bacteria and bacteriophages, cell biology, evolution, developmental biology, immunology, neurobiology, molecular medicine, and plant molecular biology.

10.011

Williams, Roger John, and Edwin M. Lansford, Jr., eds. *The encyclopedia of biochemistry*. Huntington, NY: R.E. Krieger Pub. Co., 1977. 876 p. 088275534x QP512 574.1 77-23257

(Reprint of the edition published by Reinhold, NY, 1967.) Articles discuss broad topics; alphabetic arrangement of topics in chemistry, physics, methodology, metabolism, nutrition, diseases, and disorders as they are related to biochemistry. Short biographies. Longer articles have reference lists.

NOMENCLATURE

10.012

International Union of Biochemistry, Nomenclature Committee. *Enzyme nomenclature 1984: recommendations of the Nomenclature Committee of the International Union of Biochemistry on the nomenclature and classification of enzyme-catalysed reactions.* Prepared by Edwin C. Webb. Orlando: Published for the International Union of Biochemistry by Academic Press, 1984. 646 p. 0122271629; 0122271637 pbk QP601 574.19 84-45730

Bibliography, p. 489-571. "A revision of the Recommendations (1978) of the Nomenclature Committee of IUB." Lists 2,477 enzymes with classification number, recommended name, reaction, other names, basis for classification, comments, and reference numbers. Classified arrangement; bibliography includes 4,478 references. Index to enzyme list. Updated by International Union of biochemistry and Molecular Biology. *Enzyme nomenclature 1992*: Recommendations of the *Nomenclature Committee of the International Union of Biochemistry* and *Molecular Biology on the Nomenclature and Classification of Enzymes*; see following entry.

10.013

International Union of Biochemistry and Molecular Biology. *Enzyme nomenclature 1992: recommendations* of the Nomenclature Committee of the International Union of biochemistry and Molecular Biology on the nomenclature and classification of enzymes. Prepared for NC-IUBMB by Edwin C. Webb. San Diego: Published for the IUBMB by Academic press, 1992. 862 p. 0122271645; 0122271653 pbk QP601 574.19 92-25248

A revision of the Recommendations (1984) of the NC-IUBMB.

DICTIONARIES

10.014
Dorland's illustrated medical dictionary. 28th ed. Philadelphia: W.B. Saunders Co., 1994. 1,940 p. 0721628591 (standard); 0721633777 (deluxe); 0721653235 (international) R121 610.321 78-50050

Special sections on various parts of the body; abbreviations; weights and measures; metric system; temperature scales; tables of culture media, and many more sections of interest to biochemistry students, as well as definitions of medical words.

10.015
Oxford dictionary of biochemistry and molecular biology. Managing ed., A.D. Smith [et al.]. Oxford; NY: Oxford University Press, 1997. 738 p. 0198547684 QP512 572 97-225841

Bibliographic references. Full coverage of acronyms, abbreviations, genes, and enzymes; each term is given a one-paragraph definition. Gene and protein sequences are included. Biochemical Web sites, biochemical organizations, and chemical nomenclature are in an appendix.

10.016
Stedman, Thomas Lathrop. *Stedman's medical dictionary.* 26th ed. Baltimore: Williams & Wilkins, 1995. various pagings 0683079220; 0683079352 (deluxe ed.) R121 610 94-38190

Sections offer medical etymology, blood groups, laboratory analyses, temperature scales, weights and measures, common Latin terms, Greek and Latin alphabets, and root word list as well as the vocabulary words themselves. Illustrations are included where needed for clarity.

GLOSSARIES

10.017
Bains, William. *Biotechnology from A to Z;* [foreword by G. Kirk Raab]. 2nd ed. Oxford; NY: Oxford University Press, 1998. 411 p. 0199636931 pbk TP248.16 660.6 97-36132

Glossary and dictionary of some 1,000 terms used in biotechnology.

10.018
Glick, David M. *Glossary of biochemistry and molecular biology.* rev. ed. London; Miami: Portland Press, 1997. 214 p. 1855780887 QP512 572.03 96-89356

Nearly 3,000 terms discussed in lengthy, thorough definitions, with literature references, cross-references, and acronyms included. No pronunciations.

REVIEW LITERATURE

10.019
Methods in enzymology. NY: Academic Press, 1955- . Irregular ISSN 0076-6879 QP601 574.19 54-9110

Each volume has a distinctive title, v. 8-date. Subject index, v. 1-6 issued as v. 7; v. 1-30 issued as v. 33; v. 31-32, 34-60 issued as v. 75; v. 61-74, 76-80 issued as v. 95; v. 81-94, 96-101 issued as v. 120; v. 102-119, 121-134 issued as v. 140; v. 135-139, 141-167 issued as v. 175; v. 263-264, 266-289 issued as v. 285. V. 1 (1955)-244 (1994) available on one CD-ROM. Selected issues on CD-ROM are titled *Methods in Enzymology on CD-ROM.*

10.020
Methods of biochemical analysis. NY: Wiley, 1954- . ISSN 0076-6941 QD271 543.8 54-7232

V. 1-3, 1954-56, in v. 3; v. 1-6, 1954-58, in v. 6. Beginning with v. 14, each volume contains a cumulative index to all previous volumes. Supplements accompany some volumes. D. Glick, ed. of v. 1-33; v. 34- , C.H. Suelter, ed.

HISTORY

10.021
Bowden, Mary Ellen, Amy Beth Crow, and Tracy Sullivan. *Pharmaceutical achievers.* Philadelphia, PA:

Chemical Heritage Press, 2003. 220 p. 0941901300 pbk
RM301.25 615 2001-55313

Highlights individuals who made significant achievements in pharmaceutical chemistry and biotechnology. Arranged chronologically in order of the major directions of pharmaceutical research, together with breakthroughs in treating various diseases.

10.022

Fruton, Joseph S. *A bio-bibliography for the history of the biochemical sciences since 1800.* Philadelphia: American Philosophical Society, 1982. 885 p. 0871699834
Z5524 016.57419 82-72158

Alphabetic arrangement by author; each entry has biographical and bibliographical references to books and articles. Living persons born before 1911 are included. A supplement was issued in 1986.

10.023

Fruton, Joseph S. *A skeptical biochemist.* Cambridge, MA: Harvard University Press, 1992. 330 p. 0674810775
QD415 574.19 91-29378

Fruton contemplates how biology and chemistry have come together to form biochemistry. He starts with his own ideas on the scientific method, tackles the biochemical literature, reflects on the C.P. Snow-Aldous Huxley debates, and discusses the issue of chemical classification and the development of rational nomenclature.

10.024

Hargittai, István. *Candid science II: conversations with famous biomedical scientists,* ed. by Magdolna Hargittai. London: Imperial College Press; River Edge, NJ: Distributed by World Scientific Publ, 2002. 604 p. 1860942806; 1860942881 pbk QP511.7 572 2002-510095

Hargittai interviewed three dozen bioscientists who reviewed the questions and answers to ensure an accurate record of their opinions. Yet these are not just official bio-sketches, for he used a semistructured process asking what first motivated a person to become a scientist, how their national and ethnic background influenced that process, and how they responded to world events. Most are past the age of typical retirement, many have Nobel Prizes, two are women, and all are of European ancestry.

10.025

Kohler, Robert E. *From medical chemistry to biochemistry: the making of a biomedical discipline.* Cambridge; NY: Cambridge University Press, 1982. 399 p. (Cambridge monographs on the history of medicine) 0521243122 QP511 574.19 81-10189

Discusses early specialization, institution building, and competition for funds in the rise of biochemistry to a position of importance within medical education and clinical work in the US, UK, and Germany.

10.026

Sacks, Oliver W. *Uncle Tungsten: memories of a chemical boyhood.* NY: Alfred A. Knopf, 2001. 337 p. 0375404481 RC339.52.S23 616.8 2001-33738

Sacks tells of his boyhood, growing up in Blitz-besieged Britain, and especially his "Uncle Tungsten," David Landau, the "chemistry uncle," and Abe Landau, the "physics uncle," who together ran the Tungstalite Company, which manufactured a range of incandescent and fluorescent bulbs and vacuum tubes. The two uncles supplied young Oliver with chemicals, equipment, information, and advice.

LABORATORY METHODS

10.027

Cunico, Robert L., Karen M. Gooding, and Tim Wehr. *Basic HPLC and CE of biomolecules.* Richmond, CA: Bay Bioanalytical Laboratory, 1998. 388 p. 0966322908 QP519.9.H53 547 98-70600

Thoroughly and systematically discusses separation and analysis of proteins, peptides, and polynucleotides. Covers high-performance liquid chromatography (HPLC) topics from basic theory through available chromatographies, instrumental components, and troubleshooting, sample criteria with extensive literature illustrations, data handling, and method development and validation. Capillary electrophoretic (CE) methodology is handled similarly.

10.028

Inman, Keith, and Norah Rudin. *An introduction to forensic DNA analysis.* Boca Raton, FL: CRC Press, 1997. 356 p. 0849381177 RA1057.55 614 96-38138

An updated and revised edition of Inman and Rudin's *DNA Demystified* (1994), an introductory primer on forensic DNA typing and practice. This new version has a new chapter on quality control and regulation of DNA typing practices, an expanded discussion of interpretation of DNA analytical findings, and a number of recent case studies illustrating concepts as well as introducing more esoteric DNA typing. Focuses on beginning students of DNA typing—attorneys, police officers, or forensic science students. Compact, concise style with excellent graphics. Chapter references.

10.029

Laboratory methodology in biochemistry: amino acid analysis and protein sequencing. Editors: Carlo Fini, Ardesio Floridi, Vincent N. Finelli; guest editor, Brigitte Wittman-Liebold. Boca Raton, FL: CRC Press, 1990. 263 p. 084934400x QP551 574.19 89-15747

Includes a short history of protein sequence analysis; structural determination of covalently modified peptides; amino acid analysis; structure determination of peptides and proteins; high-performance liquid chromatography; computational analysis of protein sequencing data; computers as tools in protein analysis; and other pertinent subjects.

10.030

Martin, Robin. *Gel electrophoresis: nucleic acids*. Oxford: BIOS Scientific, 1996. 175 p. 1872748287 pbk QP519.9.G42 574.87328

A useful, practical guide. Introductory chapters review nucleic acids and theoretical aspects of electrophoresis. Later chapters focus on major techniques, each introduced with a quick reference to the appropriate applications and a list of protocol sources.

10.031

Modern physical methods in biochemistry. Editors, A. Neuberger and Laurens L.M. van Deenen. Amsterdam; NY: Elsevier Science Pub. Co., 1985- . v. <1-2> (New comprehensive biochemistry, v. 11 A-B) 0444806490 (v. 1) QD415 574.19 85-4402

Surveys spectroscopic, crystallographic, chromatographic, and other physical methods in biochemistry.

TABLES OF DATA

10.032

Geigy scientific tables. 8th ed., rev. and enl. ed. [Edited by Cornelius Lentner; associate editors, Charlotte Lentner and Anthony Wink]. West Caldwell, NJ: CIBA-Geigy, 1981-1986. 4 v. 0914168509 (v. 1) QP33.5 81-70045

(Rev. ed. of *Scientific Tables*, ed. by K. Diem and C. Lentner, 7th ed., 1973) Translation of *Wissenschaftliche Tabellen*. Plastic rulers laid in; v. 2 has chart laid in. V. 1: Units of measurement, body fluids, composition of the body, nutrition; v. 2: Introduction to statistics, statistical tables, mathematical formulae; v. 3: Physical chemistry composition of blood, hematology, somatometric data; v. 4: Biochemistry, metabolism of xenobiotics, inborn errors of metabolism, pharmacogenetics and ecogenetics; v. 5: Heart and circulation; v. 6: Bacteria, fungi, protozoa, helminths.

TREATISES

10.033

Barry, J.M., and E.M. Barry. *An introduction to the structure of biological molecules*. Englewood Cliffs, NJ: Prentice-Hall, [1969]. 190 p. (Prentice-Hall biological science series) QP514 574.1 69-17705

Bridges classical organic chemistry and biochemistry and biology. Offers important structural details and procedures for separation of various groups of biologically important molecules.

10.034

Biocatalysis: from discovery to application, ed. by W.-D. Fessner. Berlin; NY: Springer, 2000. 254 p. (Springer desktop editions in chemistry) 3540669701 pbk TP248.65.E59 660.6 00-20456

Discusses state of the art of biocatalysis, from discovery of new enzymes through immobilization techniques, to their use in asymmetric synthesis of new compounds.

10.035

Comprehensive biochemistry, ed. by Marcel Florkin and Elmer H. Stotz. Amsterdam; NY: Elsevier, 1962-<1997 >. v. 1-16; 18; 18S; 19-22; 24-29; 31, pt. 3; 32-33; 34A; 35-40; in 44 v. (v. 13, 2nd ed.) 0444801510 (set) QD415 574.19 62-10359

Section 1: Physico-chemical and organic aspects of

Biological Chemistry

biochemistry (v. 1-4); section 2: Chemistry of biological compounds (v. 5-11); section 3: Biochemical reaction mechanisms (v. 12-16); section 4: Metabolism <v. 18, 18S, 19-21 >; section 5: Chemical biology <v, 22, 24-29 >; section 6: A history of biochemistry <v. 31, pt.3; 32-33; 34A; 35-40 >.

10.036

Cowan, J.A. *Inorganic biochemistry: an introduction.* 2nd ed. NY: Wiley-VCH, 1997. 440 p. 0471188956; 1560819235; 1560819359 pbk QP531 574.19 96-14100

Chapters discuss physical methods used to understand bioinorganic systems; good references and review problems.

10.037

Florkin, Marcel and Howard S. Mason, eds. *Comparative biochemistry: a comprehensive treatise.* NY: Academic Press, 1960-64. 7 v. QH345 574.192 59-132293

V. 1: Sources of free energy; v. 2: Free energy and biological function; v. 3: Constituents of life, part A; v. 4: Constituents of life, part B; v. 5: Constituents of life, part C; v. 6: Cells and organisms; v. 7: Supplementary volume.

10.038

Hammes, Gordon G. *Thermodynamics and kinetics for the biological sciences.* NY: Wiley-Interscience, 2000. 163 p. 0471374911 pbk QP517.P49 572 99-86233

Introduces biology students to some important concepts in physical chemistry used in biology. Chapters treat heat, work, and energy; entropy and free energy; chemical kinetics; application of thermodynamics; applications of kinetics; and ligand binding to macromolecules.

10.039

Jevons, Frederick Raphael. *The biochemical approach to life,* with a foreword by F. Sanger. [2nd ed.] NY: Basic Books, [1968]. 226 p. QP514 574.1 68-8988

Good elementary introduction to biochemistry, suitable for auxiliary reading.

10.040

Karlson, Peter. *Introduction to modern biochemistry,* trans. by Charles H. Doering. 4th ed. NY: Academic Press, 1975. 545 p. (Academic Press international edition) 0123997313 QH345 574.1 73-9429

Translation of *Kurzes Lehrbuch der Biochemie für Mediziner und Naturwissenschaftler.* 3rd ed., 1968; 2nd ed.,

1965. Outlines current biochemical knowledge; minimizes philosophical interpretive aspects, and emphasizes orderly listing of facts.

10.041

Klotz, Irving M. *Ligand-receptor energetics: a guide for the perplexed.* NY: Wiley, 1997. 170 p. (A Wiley-Interscience publication) 0471176265 QP517.L54 574.19 96-34518

Klotz asserts that all biological processes begin with the formation of a molecular complex from ligands and a receptor. Discusses ligand-receptor interactions from a thermodynamic point of view. Includes affinities and affinity profiles, forces of interaction, molecular scenarios, and numerical evaluation of binding constants; treats the pitfalls of data analyses. Numerous tables; four appendixes.

10.042

Lehninger, Albert L. *Lehninger principles of biochemistry.* 3rd ed. David L. Nelson, Michael M. Cox, [eds.] NY: Worth Publishers, 2000. 1 v., various pagings. 1 CD-ROM 1572591536; 1572599316; 0716738678 (CD-ROM) QD415 572 99-49137

Basic treatise on biochemistry from an author of many other biochemistry works.

10.043

Lehninger, Albert L. *Bioenergetics: the molecular basis of biological energy transformations.* 2nd ed. Menlo Park, CA: W.A. Benjamin, [1971]. 245 p. 0805360122; 0805361030 pbk QH511 574.1 71-140831

Introduction based on thermodynamic considerations; discusses source and disposition of energy allowing processes of life to occur.

10.044

Mahler, Henry R., and Eugene H. Cordes. *Biological chemistry.* 2nd ed. NY: Harper & Row, [1971]. 1,009 p. 0060441720 QP514.2 574.1 76-141169

Treats broad physical principles of bio-macromolecules, intermediary metabolism, and molecular biology.

10.045

Mathews, Christopher K., K.E. van Holde, and Kevin G. Ahern. *Biochemistry.* 3rd ed. San Francisco, CA:

Benjamin Cummings, 2000. 1,186 p. 1 CD-ROM
0805330666; 0805330674 (CD-ROM) QD415 572 99-43683

Introduces nucleic acid structure to clarify protein structure and function; emphasizes experimental roots of biochemistry and energy relationships, and explains metabolic pathways.

10.046

Mechanism and synthesis, ed. by Peter Taylor. Cambridge: Royal Society of Chemistry, 2002. 368 p. (Molecular world) 085404695x QH506

Pursues strategies for synthesizing organic compounds mainly of interest to health-related sciences. Topics include addition reactions of aldehydes and ketones; use of organometallics to form carbon-carbon bonds; and radical reactions. Retrosynthetic analysis is introduced, for developing synthesis along with biochemical pathways.

10.047

New comprehensive biochemistry. Amsterdam; NY: Elsevier/North-Holland Biomedical Press, 1981- . ISSN 0167-7306 QD415 82-3477

Editors: v. 2-date, A. Neuberger and L.L.M. van Deenen. Review articles are included that range widely through biochemical topics.

10.048

Physical chemistry of biological interfaces, ed. by Adam Baszkin and Willem Norde. NY: M. Dekker, 2000. 836 p. 0824775813 QP517.S87 570 99-51471

Treats thermodynamics, interfacial interactions, and electrical properties of interfaces and interphases; thermodynamics and kinetics of adsorption and adhesion; adsorption of biological molecules; chemistry and physics of liposomes and biomembranes; receptor-ligand interactions; interfacial enzymatic reactions; cell-cell interactions and cell adhesion; cell adhesion molecules; and adhesive strength in cell detachment.

10.049

Silva, J.J. Fraústo da, and R.J.P. Williams. *The biological chemistry of the elements: the inorganic chemistry of life*. Oxford, UK: Clarendon Press; NY: Oxford University Press, 1991. 561 p. 0198555989 QP531 574.19 91-11585

Clear diagrams and tables. Organized into three major parts; the first considers the chemical and physical properties controlling life, including the principles of uptake and the functional value of chemical elements in biological systems. The second portion treats the role of the individual chemical elements in biology, and the last section integrates the function of the elements by considering biological minerals; the elements in homeostasis, morphogenesis, and evolution; and human utilization of biological elements in the environment.

10.050

Stryer, Lubert. *Biochemistry*. 4th ed. NY: W.H. Freeman, 1995. 1,064 p. 0716720094 QP514.2 574.19 94-22832

3rd ed., 1981. New additions include metabolism and gene rearrangements. Typical biochemistry treatise.

10.051

Suckling, K.E., and C.J. Suckling. *Biological chemistry: the molecular approach to biological systems*. Cambridge; NY: Cambridge University Press, 1980. 381 p. (Cambridge texts in chemistry and biochemistry) 0521228522; 0521296781 pbk QP514.2 574.1 79-41468

Sixteen chapters treat preliminary material, reaction mechanisms, enzymes and biochemical techniques, and biological structures above the enzyme level of complexity.

10.052

Terrett, Nicholas K. *Combinatorial chemistry*. Oxford; NY: Oxford University Press, 1998. 0186 p. (Oxford chemistry masters, 2) 0198502206; 0198502192 pbk RS419 615 97-32698

Discusses practical and economic advantages of combinatorial approaches in the synthesis of libraries of compounds used in drug discovery. Combinatorial chemistry debuted in 1991, using the discovery of the Merrifield polypeptide synthesis (1963) as a start. The combinatorial method is used to make pure compounds and mixtures with biological activity. Topics covered include synthesis on resin beads, multipins (reusable polyacrylic acid-grafted polyethylene rod substrate), winks (porous polyethylene disks), and laminar solid phases; solution phase synthesis; encoded combinatorial synthesis; synthesis of compounds other than peptides (e.g., oligosaccharides, oligocarbamates, etc.); the chemistry of linking and coupling; and analysis of products by IR, NMR, and mass spectrometry. Excellent illustrations; chapter references; index; table of abbreviations.

10.053

Voet, Donald, and Judith G. Voet. *Biochemistry*. 2nd ed. NY: J. Wiley, 1995. 1,361 p. 047158651x QP514.2 574.19 94-49605

Comprehensive textbook for biochemistry majors. Contents: Introduction and background; Biomolecules; Mechanisms of enzyme action; Metabolism; The expression and transmission of genetic information.

SPECIAL TOPICS
Proteins

10.054

Adams, R.L.P., John T. Knowler, and David P. Leader. *The biochemistry of the nucleic acids*. 11th ed. London; NY: Chapman & Hall, 1992. 675 p. 0412460300; 0412399407 pbk QP620 574.87 92-213581

Chapters discuss the structure of the nucleic acids; genomes of eukaryotes, bacteria, and viruses; chromosome organization; degradation and modification of nucleic acids; metabolism of nucleotides; DNA replication, damage, repair, and recombination; arrangement of genes; RNA biosynthesis, processing of transcripts, and translation; control of transcription; and protein synthesis.

10.055

Barrett, G.C., and D.T. Elmore. *Amino acids and peptides*. Cambridge; NY: Cambridge University Press, 1998. 224 p. 0521462924; 0521468272 pbk QD431 572 97-31093

Discusses conformational aspects of amino acids and peptides; analytical procedures including chromatography and spectroscopy, immunological detection such as ELISA, and radioimmunoassay. Reactions of specific side chains and functional groups are discussed as are syntheses, biological roles, drug design, and enzyme-catalyzed peptide synthesis.

10.056

Jones, John. *Amino acid and peptide synthesis*, 2nd ed.. Oxford; NY: Oxford University Press, 2002. 92 p. (Oxford chemistry primers; 7) 0199257388 QD431 547 2002-511930

Short primer for beginners on methodology involved in peptide synthesis.

10.057

Lister, Ted. *Chemistry and the human genome*.

Cambridge: Royal Society of Chemistry, 2002. 46 p. 0854043969

Discusses the importance of the human genome, heredity, DNA sequencing, gene mapping, and the future of genome research.

10.058

Moody, Peter C.E., and Anthony J. Wilkinson. *Protein engineering*. Oxford; NY: IRL Press at Oxford University Press, 1990. 85 p. 0199531948 pbk TP248.65 660 90-7728

Reviews simply amino acid and protein structure and function; discusses mutant proteins, site-directed mutagenesis, and tailoring protein properties and functions.

10.059

Peptide biosynthesis and processing, ed. by Lloyd D. Fricker. Boca Raton, FL: CRC Press, 1991. 295 p. 084938852x QP552 574.19 91-10747

Comprehensively reviews the area, with historical perspectives to knowledge regarding specific enzymes.

10.060

Protein purification: principles, high-resolution methods, and applications, ed. by Jan-Christer Janson and Lars Rydén. 2nd ed. NY: Wiley, 1998. 0471186260 QP551 572 97-13875

Covers entire processes, from starting material, initial fractionation, and final techniques. Informational tables. Methodology of chromatography and electrophoresis.

10.061

Rhodes, Gale. *Crystallography made crystal clear: a guide for users of macromolecular models*. 2nd ed. San Diego: Academic Press, 2000. 269 p. 0125870728 QP519.9.X72 547 99-63088

A thorough, though condensed, description of computer modeling of protein crystal structures. Devoted to background material, X-ray diffraction data collection techniques, and crystal structure analysis (CSA) methods; describes such programs and instructions for importing data from existing databases, viewing molecular models, and exploring the shapes, surfaces, and packing of these molecules. Successful CSA results in a list of several hundred to several thousand atoms, each with four to nine positional and thermal parameters. Excellent illustrations; some color plates.

Lipid Biochemistry

10.062

The Lipid handbook, ed. by Frank D. Gunstone, John L. Harwood, and Fred B. Padley. London; NY: Chapman and Hall, 1986. 72, 314 p. 0412244802 QP751 574.19 85-24306

First portion is an authoritative monograph on lipid chemistry and biochemistry; the reference section following includes physical properties and literature references for about 2,000 lipids and derivatives, extracted from the *Dictionary of Organic Compounds*, 5th ed.

Foods and Nutrition

10.063

Barham, Peter. *The science of cooking*. Berlin; Heidelberg; NY: Springer, 2001. 244 p. 3540674667 TX651 641.5 00-59559

Discusses sensuous molecules; molecular gastronomy; taste and flavor; heating and eating; physical gastronomy; cooking methods and utensils; meat and poultry; fish; breads; sauces; sponge cakes; pastry; souffles; and cooking with chocolate.

10.064

Beckett, Stephen T. *The science of chocolate*. Cambridge: Royal Society of Chemistry, 2000. 175 p. 0854046003 TP640 664.5

Describes the history of chocolate, ingredients and processing techniques, monitoring and controlling of chocolate production, and packaging of chocolate products.

10.065

Bender, David A. *Nutritional biochemistry of the vitamins*. Cambridge; NY: Cambridge University Press, 1992. 431 p. 0521381444 QP771 612.3 91-34082

Reviews and discusses vitamin A: retinol and beta-carotene; vitamin D; vitamin E: tocopherols and tocotrienols; vitamin K; vitamin B1: thiamin; vitamin B2: riboflavin; niacin; vitamin B6; folic acid and other pterins and vitamin B12; biotin (vitamin H); pantothenic acid (vitamin B5); ascorbic acid (vitamin C); compounds of doubtful vitamin status: taurine, L-carnitine, choline, and inositol.

10.066

Berdanier, Carolyn D. *CRC desk reference for nutrition;* with contributions from Anne Dattilo and Wilhelmine P.H.G. Verboeket-van de Venne. Boca Raton, FL: CRC Press, 1998. 358 p. (CRC desk reference series) 0849396824 QP141 613.2 97-43853

Includes common medical terms and descriptions of biochemical pathways and physiological processes. Terms are listed alphabetically; figures and diagrams are included. Answers questions relating to nutrition as applied to human health and well-being. Recent developments in physiology, biochemistry, molecular biology, and pathology are utilized to explain how nutrition science is related to normal body growth and function. An essential reference.

10.067

Brody, Tom. *Nutritional biochemistry*. 2nd ed. San Diego: Academic Press, 1999. 1,006 p. 0121348369 QP141 612.3 98-40384

Explanations are illustrated with carefully selected reactions, diagrams, and structures and examples. A fully referenced, up-to-date work; the last chapter treats diet and cancer.

10.068

Chemical and functional properties of food components, ed. by Zdzisław E. Sikorski. Lancaster, PA: Technomic Pub. Co., 1997. 293 p. 1566764645 TX545 664 96-61440

Discusses basics as well as deeper topics; 12 chapters overview major food component classes: water, carbohydrates, proteins, lipids, minerals, flavors, and colors. The remaining chapters deal with food quality, rheological properties, food safety, and mutagenic/carcinogenic components in food. Chapter references; tables and figures; numerous chemical structural formulas.

10.069

The Concise encyclopedia of foods & nutrition, by Audrey H. Ensminger [et al.]. Boca Raton, FL: CRC Press, 1995. 1,178 p. 0849344557 TX349 613.2 94-3000

Foods, health, and nutrition are discussed, from vitamins to vitality, from fiber to fad foods, from malnutrition to minerals, from disease to diet.

10.070

Coultate, T.P. *Food: the chemistry of its components*. 4th ed. Cambridge: Royal Society of Chemistry, 2002. 432

p. (RSC paperbacks) 0854046151 TX551 641.300154

Discusses carbohydrates, fats, proteins, minerals, and water; pigments, flavors, vitamins, and preservatives; modified starches, naturally occurring antioxidants, health benefits of cruciferous vegetables, and production of glucose syrup.

10.071

Davies, Michael B., John A. Austin, and David A. Partridge. *Vitamin C: its chemistry and biochemistry*. Cambridge: Royal Society of Chemistry, 1991. 154 p. (Royal Society of chemistry paperbacks) 0851863337 pbk QP772 574.19 91-93545

Contents: Introduction; History of vitamin C and its role in the prevention and cure of scurvy; Discovery and structure of vitamin C; Synthesis, manufacture, and further chemistry of vitamin C; Biochemistry of vitamin C; Medical aspects of vitamin C; Inorganic and analytical aspects of vitamin C chemistry.

10.072

Encyclopedia of vitamins, minerals, and supplements. Tova Navarra and Myron A. Lipkowitz. NY: Facts on File, 1996. 281 p. 0816031835; 0816032416 pbk QP771 612.3 95-12645

Covers mostly herbal and other supplements used in alternative or traditional medicine as practiced by Native Americans, Chinese, and other ethnic groups.

10.073

Nutrition, an integrated approach, [ed. by] Ruth L. Pike and Myrtle L. Brown. 3rd ed. NY: Wiley, 1984. 1,068 p. 0471090042 QP141 613.2 83-16766

Discusses nutrients; historical perspective; carbohydrates, lipids, proteins, nucleotides, and nucleic acids; water-soluble vitamins; fat-soluble vitamins and other vitamins; minerals and water; physiological aspects of nutrition; digestion and absorption; transport and exchange; the cell; orientation to cellular nutrition; the plasma membrane; the nucleus; the cytoplasmic matrix and endoplasmic reticulum; the Golgi apparatus; the mitochondria; the lysosomes and microbodies; specialized cells; hepatocytes; erythrocytes; bone, muscle, nerve, and adipose cells; the complex organism; cellular growth; body composition; determination of nutrient needs; energy, protein, minerals; determination of nutrient needs: vitamins; dietary standards; nutrition surveys.

10.074

Nutrition: chemistry and biology. 2nd ed. Julian E. Spallholz, L. Mallory Boylan, and Judy A. Driskell. Boca Raton, FL: CRC Press, 1998. 352 p. (CRC series in modern nutrition) 0849385040 QP141 572 98-34621

Includes food sources of nutrients; nutrient toxicities and interactions; sugar and fat substitutes; nutritional therapy applications; effects of cooking methods on nutrients; dietary intakes of nutrients in the U.S.; dietary recommendations; and free radicals and antioxidants.

10.075

Somer, Elizabeth. *The essential guide to vitamins and minerals*. 2nd ed. NY: HarperPerennial, 1995. 449 p. 0062715941; 0062733451 pbk QP771 641.1 91-55390 (1st ed.)

Nutrition as a way of life; vitamins, minerals, and the body; the vitamins; the minerals; vitamins, minerals, and disease; how medications, alcohol, and tobacco affect vitamin and mineral status; vitamins, minerals, and food; understanding supplements.

Medical Biochemistry and Pharmacology

10.076

Bhagavan, N.V. *Medical biochemistry*. Boston: Jones and Bartlett Publishers, 1992. 980 p. (The Jones and Bartlett series in biology) 0867200308 QP514.2 612 91-7015

(Rev. ed. of *Biochemistry*, 2nd ed., 1978) Basic biochemistry is presented in a context that stresses its significance in the understanding of human disease and its treatment.

10.077

Cannon, Joseph G. *Pharmacology for chemists*. Washington, DC: American Chemical Society; NY: Oxford University Press, 1999. 373 p. (ACS professional reference book) 0841235244 RM300 615 98-25069

Three sections include an introduction to various chemical and biological topics; the central and peripheral nervous system and drugs used in these areas; and the cardiovascular and allergic reaction systems. Diagrams and flowcharts. Designed to educate chemists about the pharmacology of drugs they know chemically, and can serve as a chemistry primer for health-care workers.

10.078

Drugs. Ram N. Gupta, ed.; Irving Sunshine, consulting ed. Boca Raton, FL: CRC Press, 1981-<1989>. v. <1, 3-6 >. (CRC handbook of chromatography) 0849330300 (set) RS189 615 81-10157

Drugs are arranged alphabetically according to generic names; synonyms and proprietary names are found in the appendix and in the index.

10.079

Gringauz, Alex. *Introduction to medicinal chemistry: how drugs act and why*. NY: Wiley-VCH, 1997. 721 p. 0471185450 RS403 615 95-49331

Explains drugs by combining necessary biological/physiological concepts with chemistry. Discusses shortcomings and hazards of drugs used therapeutically, mechanisms of action, and stability in the body and on the shelf. Anticancer drugs, analgetics, antimicrobials, steroids, and cardiovascular drugs are discussed.

10.080

Hansch, Corwin. *Comprehensive medicinal chemistry*. NY: Pergamon, 1990. 6 v. 0080325300 (set); 0080370578 (v. 1); 0080370586 (v. 2); 0080370594 (v. 3); 0080370608 (v. 4); 0080370616 (v. 5); 0080370624 (v. 6) RS403 615.19 89-5863

Discusses the rational design, mechanistic study, and therapeutic application of chemical compounds. V. 1: General principles; v. 2: Enzymes and other molecular targets; v. 3: Membranes and receptors; v. 4: Quantitative drug design; v. 5: Biopharmaceutics; v. 6: Cumulative index, Drug compendium.

10.081

Hardie, D.G. *Biochemical messengers: hormones, neurotransmitters, and growth factors*. London; NY: Chapman & Hall, 1991. 311 p. 041230340x; 0412303507 pbk QP517 574.87 91-8226

Discusses the task of describing and integrating the various mechanisms used for intracellular and extracellular signaling.

10.082

Repic, Oljan. *Principles of process research and chemical development in the pharmaceutical industry*. NY: Wiley, 1998. 213 p. (A Wiley-Interscience publication) 0471165166 RS403 615 97-13904

A universal tool for synthetic chemists, wherever they are; offers on-the-job training for economic and environmental issues, lab safety, and plant design, while explaining why scaling up is more than multiplying a recipe by a factor. Chapters cover impurities, rational drug design and chirality, strategies for synthesis, the role of isotopes, licensing, and the future. References current through 1996.

Toxicology and Environmental Concerns

10.083

Landis, Wayne G., and Ming-Ho Yu. *Introduction to environmental toxicology: impacts of chemicals upon ecological systems*. 2nd ed. Boca Raton, FL: Lewis Publishers, 1999. 390 p. 1566702658 QH545 571.9 97-50324

Emphasizes an ecosystem approach, and covers topics such as routes of exposure, modes of action, factors affecting toxicity, biotransformations, inorganic pollutants, surveys of laboratory methods for toxicity testing, and ecological risk assessment at ecological levels of organization. Chapter questions and references.

Plant and Animal Biochemistry

10.084

Britton, George. *The biochemistry of natural pigments*. Cambridge; NY: Cambridge University Press, 1983. 366 p. (Cambridge texts in chemistry and biochemistry) 0521248922 QD441 82-9512

Treats the chemistry and biochemistry of natural pigments, and the functional role of these pigments in nature. Includes carotenoids, quinones, flavonoids, tetrapyrroles, other nonpolymeric N-heterocyclic pigments, and the melanins. Two chapters discuss light and color and the importance of color in nature.

10.085

The carbohydrates: chemistry and biochemistry, ed. by Ward Pigman, Derek Horton; assistant editor, Anthony Herp. 2nd ed. NY: Academic Press, 1970-1980. 2 v. in 4 0125563027 QD321 547 68-26647

Long, encyclopedic chapters written by experts.

10.086

David, Serge. *The molecular and supramolecular chemistry of carbohydrates: chemical introduction to the glycosciences*, transl. by Rosemary Green Beau. Oxford;

NY: Oxford University Press, 1997. 320 p. 0198500475; 0198500467 pbk QP701 572 97-18604

Discusses classical carbohydrate chemistry topics (configurations, conformations, chemical reactions, nomenclatures, and so forth) and biological applications (associations, glycoconjugates, antigens and antibodies, blood group antigens, ligands to DNA, and more).

10.087

Milgrom, Lionel R. *The colours of life: an introduction to the chemistry of porphyrins and related compounds.* Oxford; NY: Oxford University Press, 1997. 249 p. 0198553803; 0198559623 pbk QD441 547 96-27452

Chapters discuss classes of porphyrins, naming systems and, briefly, biological functions of porphyrins; prebiotic synthesis, biosynthesis, and chemical synthesis; review of essential concepts (orbital theory, electrophylic aromatic substitution, ultraviolet spectroscopy, etc.) and a simple treatment of the mechanism of oxygen production in chlorophyll-moderated photosynthesis; the role of hemoglobin and myoglobin in oxygen uptake, transport, and release; catabolism of biologically occurring porphyrins including a discussion of petroporphyrins; porphyrin pathologies, e.g., methemoglobinemia, sickle-cell anemia, porphyria; and commercial and therapeutic applications (actual and potential), e.g., pigments, molecular conductors, cancer therapy. Clear structures and diagrams; numerous chapter references.

10.088

Robyt, John F. *Essentials of carbohydrate chemistry.* NY: Springer, 1998. 399 p. (Springer advanced texts in chemistry) 0387949518 QD321 572 97-19019

This book provides a good overview of the field from both historical and technical perspectives. In addition to topics such as chemical transformations and modifications, structure and function, and biological synthesis and degradation, topics such as the development of carbohydrate chemistry and sweetness are also included.

10.089

Stick, Robert V. *Carbohydrates: the sweet molecules of life.* San Diego, CA; London: Academic, 2001. 256 p. 0126709602 QD321 547.78

An excellent resource for anyone who has completed a one-year organic chemistry course. This book is an excellent account of the science, starting with the works of the pioneers

and continuing through the significant results published within the last year. Coverage includes structure and conformation determination, protecting group chemistry, reactions of monosaccharides, glycosidic bond formation, and oligosaccharide synthesis. There are brief reviews of disaccharides, oligosaccharides, polysaccharides, glycoconjugates, glycobiology, and carbohydrate-based vaccines.

Steroids

10.090

Richards, John H., and James B. Hendrickson. *The biosynthesis of steroids, terpenes, and acetogenins.* NY: W.A. Benjamin, 1964. 416 p. (Frontiers in chemistry) QD415 574.192 64-21233

Presents current theories of synthesis of steroids, terpenes, and related compounds of plant and animal origin. Footnote references.

LABORATORY SAFETY

10.091

Picot, André, and Philippe Grenouillet. *Safety in the chemistry and biochemistry laboratory.* Andrew T. Prokopetz and Douglas B. Walters, editors of the English-language ed. Translator, Robert H. Dodd; foreword by Sir Derek Barton. NY: VCH, 1995. 318 p. 1560810408 QD63.5 542 94-38824

The authors review the full gamut of possible risks, ranging from general hazard types such as explosion, flammability, toxicity, biological hazards, and radiation, to information about specific compounds. Readable translation with balanced coverage of English-language references and practices. Comprehensive bibliography; very useful section on neutralization and destruction of chemical wastes.

PROBLEM MANUALS

10.092

Hames, B.D., and N.M. Hooper. *Instant notes in biochemistry.* 2nd ed. Oxford: BIOS Scientific Publishers, 2000. 422 p. (The Instant notes series) 1859961428 pbk 572

Concise but comprehensive coverage of biochemistry featuring cell organization, amino acids and proteins, enzymes, antibodies, membranes, DNA structure and replication, RNA synthesis and processing, protein synthesis, recombinant DNA technology, carbohydrate metabolism, lipid metabolism, respiration and energy, nitrogen metabolism, and cell specialization.

10.093

Schaum's outline of theory and problems of biochemistry; Philip W. Kuchel ... [et al.]. 2nd ed. NY: McGraw-Hill, 1998. 559 p. (Schaum's outline series) 0070361495 pbk QP518.3 572 97-23525

Chapters treat cell ultrastructure, carbohydrates, amino acids and peptides, proteins and their supramolecular structure, lipids, membranes, transport, signaling, nucleic acids, enzyme catalysis and kinetics, metabolism, carbohydrate metabolism, citric acid cycle, lipid metabolism, oxidative phosphorylation, nitrogen metabolism, replication and maintenance of genetic material, and gene expression and protein synthesis.

Chapter 11

Internet Resources in Chemistry

This final chapter introduces the vast chemical information resources available via the Internet. Up to this time, we have noted various electronic materials useful in the profession of chemistry, such as CD-ROMs, together with other related materials. Here we will not discuss CD-ROMs and other types of stand-alone materials because they have already been considered.

The Internet (also named the World Wide Web, though this is but a portion of the Internet) had its beginnings with the Army's Advanced Research Projects Agency (ARPA) research efforts in the 1960s. From that agency's work was developed a network in which there would be a continuous flow of material, without interruption, because all users were linked to all other users, and the information flowed through a network rather than from one to another in a chain-like fashion. From ARPA efforts the Internet as a public utility became a reality in the early '90s. Since then, refinement after refinement has made the Internet a most useful entity, from every standpoint—entertainment, utility, information, and resource storage.

Here are some of the most useful chemistry-oriented Web sites (and a few print sources) that are useful for student, faculty, and industrial use, and have appeared to be the most stable (read: not readily disappearing) sites of all.

GUIDES TO THE INTERNET

Printed materials

11.001

Brecher, J.S. "The *ChemFinder WebServer*: indexing chemical data on the Internet." *Chimia* **52**, 658-663 (1998); available from http://www.chemfinder.com/chimia.html

The Internet offers a wealth of data, but chemical information provides unique challenges for searching. This article summarizes general indexing techniques and explains why those techniques are unsuited to chemical data. The *ChemFinder WebServer*, a WWW-based chemical index, will help to overcome the weaknesses of other Internet search engines.

Internet resources

11.002

Chemdex.org

http://www.chemdex.org/ or:

http://www2.shef.ac.uk/chemistry/chemdex/

Chemdex.org maintains a directory of chemistry on the Internet, and originates from Sheffield University in England. There are links to universities, institutes, government organizations, companies, institutions, divisions of chemistry, databases, communication networks, software packages, and World Wide Web chemistry developments. Some miscellaneous links are provided.

11.003

ChemFinder.com

http://www.chemfinder.com/

This commercial site, designed to find information on chemicals, features indexed sites, a glossary, materials for downloading, subscription information, and other items of interest. Feedback and other user-related functions are available. Searching is by name, CASRN, structure, and so forth.

11.004

Chemie.de

http://www.chemie.de/

This Web site from Germany has links to various items of interest to chemists: a job exchange, search engine, chemistry departments, software, chemistry forum, buyers' guide, conferences, chemistry toolbox, newsletter, and information.

11.005

ChemIndustry.com

http://www.chemindustry.com/ or http://www.neis.com/

This site, billed as "the leading comprehensive directory and search engine for chemical and related professionals," is industrially oriented and offers links to industry sectors, equipment and software, chemical resources, portals and news, industry services, career and community, chemical technology, organizations, events, and academic institutes.

11.006

ChemWeb.com

http://www.chemweb.com/

This Web site contains a library of leading chemical journals; chemical databases; *The Alchemist*, an online journal of latest chemical news; a shopping mall; worldwide job exchange; conference diary; *ChemDex Plus*, a searchable and reviewed database of chemistry resources on the Internet; and ACD, available chemicals directory.

11.007

Huber, Charles F. "Chemistry resources on the Internet." *Issues in Science and Technology Librarianship*, Winter 1998, http://www.library.ucsb.edu/istl/98-winter/interne1.html

Huber discusses resources available via the Internet; he includes data, materials safety data sheets, chemistry departments, super sites, awards and prizes, professional societies, conferences, meetings, symposia, commercial information providers, electronic journals, chemical suppliers, chemical equipment suppliers, and software vendors. All the sources are Web sites.

11.008

Links for chemists

http://www.liv.ac.uk/Chemistry/Links/links.html

This site from the University of Liverpool is an index to various links of interest to chemists (some 8,075 as of 2001). One may do keyword searching of the index; there is linking to international chemistry department Web sites and those in the UK. New link listings are available by month of listing.

11.009

National science digital library

http://nsdl.org/

This site serves as an educational resource for science, technology, engineering, and mathematics. Funded by the National Science Foundation. Much of the material is appropriate for students training for teacher education.

11.010

Yahoo! Science: Chemistry

http://dir.yahoo.com/Science/Chemistry/

Yahoo! is actually a utility rather than a Web site; it functions as a directory or pointing device rather than as a conventional Internet site. Under the Chemistry rubric, there are listed nearly 40 different possible pathways to take, with numbers of hits for each. There are other items that Yahoo! considers features of the day/week/month, and are listed below the listing. Definitely the place to start a search! (Any unfulfilled search on Yahoo! will default to Alta Vista, another search engine for the Internet.)

CHEMICAL INFORMATION RESOURCES

11.011

American Chemical Society publications

http://pubs.acs.org/

The ACS publications Web site contains announcements of new articles from various ACS journals, updated whenever a new issue is published. Also available are links to ACS journals and magazines, directories and buyers' guides, advertised products, jobs, library links, and other items of interest. Links take users to *ChemPort*, Chemical Abstracts Service, and ACS's *ChemCenter*.

11.012

Chemical Abstracts Service (CAS)

http://www.cas.org/

Here is *Chemical Abstracts*, without which no chemist can keep house! CA is searchable via this Web site using two services: STN Easy, and STN International; SciFinder and SciFinder Scholar are featured, two of CA's products that offer a search interface for CA. The STN system offers 200 scientific and technical databases from around the world. All CAS products are fee-based; users should contact CAS for further information and fee schedules.

11.013

CHEMINFO: chemical information sources from Indiana University

http://www.indiana.edu/~cheminfo/

This monumental Web site offers just about everything anyone would want to see in the chemical information arena. Site author Gary Wiggins includes his entire chem lit curriculum, with listings of all books and other materials; selected Internet resources; lecture notes in chemical information; the Clearinghouse for Chemical Information Instructional Materials; chemical information sources discussion list archives; the Indiana University Chemistry Library; biotechnology resources; and the Indiana University Molecular Structure Center. Each of these sections contains a huge number of other resources, which in turn may lead to still others, as well as links to other Web sites.

11.014

ChemPort

http://www.chemport.org/

ChemPort offers direct links from various search services for obtaining full-text copies of articles for a fee. *ChemPort* is linked to STN Easy, STN Express with Discover, SciFinder, and SciFinder Scholar (see *Chemical Abstracts* Web site for information on these products).

11.015

Crystallography online

http://www.iucr.org/cww-top/crystal.index.html

The International Union of Crystallography maintains this site; it includes structures of minerals, rocks, physics of X-rays and magnetic materials, shapes and sizes of molecules, and molecular biology structures of proteins and nucleic acids. Several important links to directories and online journals enhance this site.

11.016

NIST chemistry Webbook

http://webbook.nist.gov/chemistry/

With this newest NIST Standard Reference Database Number 69, July 2001 release, this Web site has been updated and contains information of very high quality and reliability, and is distinctive. Like other material published by NIST (formerly the National Bureau of Standards), this database is authoritative and represents critical evaluation of relevant data by experts. Coverage is broad, with data on some 40,000 total species including thermochemical data on more than 6,500 compounds, thermodynamic data for some 29,200 total species, gas phase ion data for 16,800 total species, spectral data for about 21,700 species, and energy-related data for about 4,700 materials. Since several million compounds are known to chemists, there is room for growth. NIST data can be searched very flexibly by name, chemical formula, CASRN, molecular weight, ionization energy, or proton affinity. Thorough documentation is provided, including publications organized by author as well as substance. A guide is provided for citation of data from this database.

11.017

Table of isotopes

http://ie.lbl.gov/toi.htm

Kept up-to-date by Berkeley's Lawrence National Radiation Laboratory. Provides very detailed data on nuclear structures in atoms, energies, and decay products.

RESOURCES FOR EDUCATORS IN CHEMISTRY

Many of the Internet sites seen contain materials for courses, entire curricula, or discussions of various portions of the chemical curricula.

11.018

Index to chemical education resources

http://www.umsl.edu/~chemist/books/texts.html

Part of the *Journal of Chemical Education* Website, this is a version of the once-annual book and journal list included in the *Journal of Chemical Education*. Books are listed chronologically, from newest to oldest; within each year, listings are alphabetical by first author. Book reviews are referenced and hyperlinked if from *Journal of Chemical Education*. Software products are listed under the publisher/distributor name. Hyperlinks are provided to the software producers.

11.019

Journal of Chemical Education

http://jchemed.chem.wisc.edu/

This journal Web site contains an index to the *Journal*, publications, book information, and software, all available from the journal's publisher.

11.020

The "Virtual" chemistry center

http://www-sci.lib.uci.edu/HSG/GradChemistry.html>

Martindale's Virtual Chemistry Center is distinctive in several respects; it provides an unusually comprehensive collection of links and is quite eclectic in its selection of sites related to chemistry. The site is updated on virtually a daily basis; it has numerous links and excellent content. *The "Virtual" Chemistry Center* is organized into several broad main categories; these include Measurement; Periodic Tables; Chemicals & Biochemicals: Major Databases; Chemicals & Biochemicals: Specialized Databases; Chemistry Overview; and Chemistry Courses, Textbooks & Databases. There are large numbers of subcategories and topics; a few examples include Hazardous Chemicals, Patent Search, Safety Manuals & Guidelines, Computer Chip Chemistry, Art & Animation, What's New in Chemistry, Interactive Laboratory Experiments, and Nanotechnology.

11.021

WebElements

http://www.webelements.com

This interactive Web site is based on the periodic table of elements, with a wealth of information regarding the atomic elements. It is a delightful blend of content, intelligent design, and graphical features. Information related to each element is divided into sections and includes background, crystallography, isotope, and spectroscopy information as well as chemical, electronic, biological, and geological data. The data quality and reliability of the content seem very good. Three-dimensional crystal structures are provided for most elements using graphics, PDB (Protein Data Bank) Web viewers (atomic coordinate files), and virtual reality Web plug-ins. The element names, pronounced in English, are also included as sound files.

SOCIETIES AND ORGANIZATIONS

11.022

Chemistry.org

http://www.chemistry.org

Chemistry.org links to major Web resources developed by the ACS Publications Division and the Chemical Abstracts Service, as well as to Internet sites of scientific associations, publishers, vendors of information services, universities, and other content providers. The stature of the ACS among scientific associations makes this site particularly valuable as a virtual library of chemistry resources that have been validated by experts. Students at all levels will find numerical and factual sources of information from the periodic table through directories of graduate educational opportunities. The "Science Smorgasbord" section offers a collection that should assist in completing course assignments; it will also entertain and inspire students by providing a rich sampling of developments in research. There are also links to electronic publications, some full-text and others as tables of contents or summaries. Other virtual libraries of chemistry-related resources are linked from *Chemistry.org*, but ACS's imprimatur makes this site the preferred starting point for an Internet search. A few connections are password-protected and link to fee-based services such as STN (Scientific and Technical Network International) and its offerings of scientific databases.

11.023

Chemsoc

http://www.chemsoc.org/

This Web site is hosted by the Royal Society of Chemistry and serves as a starting "gate" for library information, educational resources, societies directory, conferences and events, Chembytes news and magazine, and a links page, called "Chemistry Resource Locator." There are some 140 chemically related, international societies listed, some with e-mail addresses, and there is a search function for searching the societies' Web sites, together and singly.

Journals in Chemistry

Those charged with assembling a collection of relevant journal titles should consult the *Library Guidelines for ACS Approved Programs* available on the American Chemical Society Web site at http://www.chemistry.org/portal/a/c/s/1/ acsdisplay.html?DOC=education/cpt/library.html, which contains a list of key journals. The most current list as of press time is reprinted below. Titles published by the American Chemical Society are recommended for any basic chemistry journal collection.

American Chemical Society, Committee on Professional Training
Journal List for Undergraduate Programs (as of January 2004)

General Content

Accounts of Chemical Research
Angewandte Chemie International Edition
Chemical Communications
Chemical Reviews
Chemical Society Reviews
Journal of Chemical Education
Journal of the American Chemical Society
Nature
Proceedings of the National Academy of Sciences
Science

Topical

Highly Recommended

Analytical Chemistry
Biochemistry
Chemistry of Materials
Dalton Transactions
Environmental Science & Technology
Faraday Discussions
Inorganic Chemistry
Journal of Biological Chemistry
Journal of Chemical Physics
The Journal of Organic Chemistry
The Journal of Physical Chemistry A
The Journal of Physical Chemistry B
Langmuir
Macromolecules
Nature-Structural Biology
Organic Letters
Organometallics
Organic and Biomolecular Chemistry (formerly Perkin Transactions 1 and 2)

Also Recommended

Biochemical Journal
Bioconjugate Chemistry
Bioorganic Chemistry
Canadian Journal of Chemistry
Chemical Physics Letters
Chemistry-A European Journal
Chemistry and Biology
Chemistry Letters (Japan)
European Journal of Biochemistry
European Journal of Organic Chemistry
Helvetica Chimica Acta
Industrial & Engineering Chemistry Research
Inorganica Chimica Acta
Journal of Biological Inorganic Chemistry

Journal of Catalysis
Journal of Chemical Information & Computer Science
Journal of Chromatography
Journal of Coordination Chemistry
Journal of Electroanalytical Chemistry
Journal of Medicinal Chemistry
Journal of Molecular Biology
Journal of Organometallic Chemistry
Journal of Polymer Science
Magnetic Resonance in Chemistry
New Journal of Chemistry
Physical Chemistry Chemical Physics
Pure and Applied Chemistry
Spectrochimica Acta
Tetrahedron
Tetrahedron Letters
Trends in Biochemical Sciences

The Society notes: "In addition to the required journals, undergraduates may benefit from browsing a collection including some of the following: Chemical and Engineering News, Chemistry World (formerly Chemistry in Britain), Scientific American, and Chemtracts."

Other Resources for Journals

In addition to the above information, any number of Web sites provide useful lists of journals; some contain hotlinks to publishers or other sources of further information. Using the top Internet search engines will reveal many lists offered by various colleges and universities throughout the world. The following sites are noted for their large collections of journal titles:

Cambridge University (UK) Chemistry Journals
http://www.ch.cam.ac.uk/c2k/cj/
This site claims to have "one of the world's most comprehensive and up-to-date lists of Internet-linked chemistry-related journals." The list runs to 21 pages. The list is also available sorted by publisher.

The Royal Society of Chemistry (UK)
http://rsc.org/is/journals/current/ejs.htm
The Royal Society maintains a list of online journals, both from the RSC and from other publishers, worldwide. It also lists online journals that are available only electronically or now incorporated under new titles.

Chemical Abstracts Service (CAS)
http://www.cas.org/sent.html
Finally, CAS has published a core list of journals covered in CAplus, which contains the short title, complete title, CODEN, and publication frequency for the key/core journals abstracted in *Chemical Abstracts* and available via an electronic version on STN. As the site indicates, this is not the full list of publications covered by CAS.

■ INDEX

178

J

K

Q

R

U

V

W

X - Z

Printed in the United States
48031LVS00002B/257-272

9 780838 983089